Ignacio Corazza

Use of draining concrete

Ignacio Corazza

Use of draining concrete

for stormwater overflow receiving structures

ScienciaScripts

Cover image: www.ingimage.com

This book is a translation from the original published under ISBN 978-620-2-15342-3.

Publisher:
Sciencia Scripts
is a trademark of
Dodo Books Indian Ocean Ltd. and OmniScriptum S.R.L publishing group

120 High Road, East Finchley, London, N2 9ED, United Kingdom
Str. Armeneasca 28/1, office 1, Chisinau MD-2012, Republic of Moldova, Europe
Printed at: see last page
ISBN: 978-620-7-91258-2

Contents

Chapter I

Introduction to the problem

The population increase is a phenomenon of global scale, currently the population growth rate is 0.82% per year for the world population and 0.5% per year for Argentina (both values for the year 2021) according to the United Nations in its report of July 2022 [1].

This is accompanied by the phenomenon of urbanisation, as people move to cities seeking easier access to goods and services, with the world's urban population expected to grow from 55.3% in 2018 to 68.4% by 2050 [2]. The demand for housing and other buildings generates an increase in the price of land in urban areas, which is why they seek to increase their profitability, creating housing towers that increase population density. This process demands infrastructure for the movement of people, goods, etc., generating, together with housing, businesses, public buildings and other buildings, an increase in the waterproofed surface. This increase in the impermeable surface area in cities increases the probability of flooding caused by overflows of the public sewerage system, as it reduces the time of concentration of the basins, modifying the runoff hydrograph and increasing the flow to be disbursed, which must be conducted through existing conduits, deteriorated by the passage of time, clogged by waste and designed for lower peak flows [3, 4].

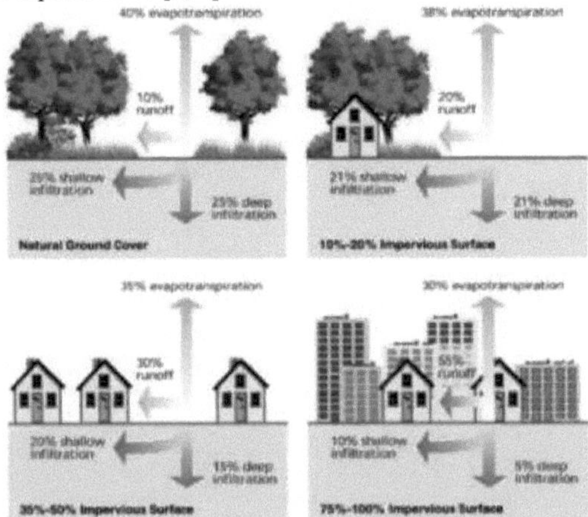

Figure N° 1. Comparative diagram of the hydrological cycle [5].

As can be seen in figure 1, the percentage of precipitation that becomes surface runoff increases in relation to the impermeabilisation of the surface. When the entire ground is covered by natural soil, only 10% of precipitation becomes surface runoff, whereas when 75 to 100% of the ground is impermeable, this value rises to 55%. Another consequence of urbanisation is the presence of waste in stormwater; cities demand a large amount of goods and services in small areas, where if there is no comprehensive and efficient programme for the management of urban waste, there can be a high concentration of pollutants in the streets and pavements, which are carried by rainwater, conducted through stormwater pipes and then deposited in natural receiving watercourses, producing negative impacts on infrastructure, the environment and the health of the inhabitants [6].

Damage to the infrastructure includes deterioration of the surface due to friction with the objects dragged along and chemical reactions with pollutants. It also increases the likelihood of obstructions that hinder or impede the flow of water, leading to overflows [7].

Environmental impacts include damage to the ecosystem, affecting the flora and fauna present in the watercourse receiving the waste, which may result in the natural relocation of species, overpopulation of certain specimens, elimination of others, etc. This also entails risks of contamination of food obtained from the environment, such as fish, increasing the cost of quality control and cleaning, or even the impossibility of obtaining safe food, forcing citizens to look for new sources, which implies an increase in the cost of transporting and storing food.

Another environmental impact is on water quality, which in addition to being detrimental to the ecosystem implies health risks for those who come into contact with the contaminated water, such as those who live in the areas surrounding the course, or who carry out recreational, sporting or work activities in it, leading to an increase in mortality and/or illnesses. This also implies an increase in the cost of making the water potable for human consumption or in the processes necessary to make it suitable for industrial use.

All the increases in costs associated with each of the activities described above in turn produce a greater consumption of goods and services, thus increasing the energy used, producing greenhouse gases, aggravating one of the origins of the problem, which is climate change. As can be seen, it is a cycle that, in the absence of any intervention, would continue to feed back over time, aggravating the consequences described above and/or creating others.

1. Treatment of the problem

Below are different alternatives adopted in different areas of the world in order to prevent overflows, improve the quality of water discharged through rainwater pipes, improve the process of recharging aquifers, etc. As humans use water, it becomes contaminated with oils, sediments, detergents, heavy metals, fertilisers, pathogens, pesticides and waste, among others. This decrease in water quality becomes a public health problem and a deterioration of the environment. This also occurs in rainwater, which absorbs impurities present in the air and carries away those found on city surfaces (roofs, pavements, streets, etc.). This is further aggravated by the removal of barriers that can stop pollutants (e.g. the natural filter provided by some soils) and in most cases they travel to large water bodies without any treatment.

Among the new strategies dedicated to improving the functioning and sustainable urban development of cities are the so-called "Sustainable Urban Drainage Systems" (SUDS). These emerged around the 1980s as an alternative to conventional stormwater drainage. SUDS minimise impacts on the hydrological cycle through source control practices, reducing the input of pollutants into stormwater runoff, and also make use of practices that encourage on-site management of rainfall, temporary treatments, detention and infiltration. In addition to the benefit of flood control, they improve the quality of water delivered, encourage aquifer recharge through percolation processes and provide benefits to wildlife [8].

These sustainable urban drainage systems are designed to emulate, as closely as possible, the natural regime of the hydrological cycle present in the area to be developed once the urbanisation process has taken place. Sustainable urban drainage systems involve not only structural but also non-structural measures.

Non-structural measures seek to raise awareness of the problem among the population, encouraging better water use, better waste management, generating control measures in the sources that generate pollution, among other aspects. Among the main ones are:

- Awareness-raising campaigns on the problem, seeking a change in citizens' habits.

- Identification of main pollutants and their sources.
- Generation of standards for industries and other centralised sources of pollution.
- Planning impervious surfaces to reduce runoff.
- Encouraging frequent cleaning of impervious surfaces and drains to reduce the accumulation of pollutants.
- Control of herbicide and fungicide application in parks and gardens.
- Prevent sediment carry-over from on-site areas.
- Provide spill clean-up procedures using dry techniques.
- Avoid contact of rainwater with pollutants as far as possible.
- Separate stormwater from wastewater.
- Store and reuse rainwater.

Structural measures consist of civil structures of low environmental impact that contribute to the reduction of impervious areas, minimising surface runoff through infiltration, temporary retention and percolation, and favour the improvement of water quality. Also included are works for the temporary storage of rainwater.

- Natural surfaces: these are densely vegetated strips with a gentle slope, created to improve water quality through the filtration process, favouring infiltration into the groundwater table and reducing surface runoff. This can be seen in the following image.

Figure N° 2: Vegetated channel: (Source: Thomas Engineering PA: http://www.thomasengineeringpa. com).

- Permeable pavements: on surfaces where it is necessary to replace the natural soil for vehicle traffic, they can be covered with concrete or asphalt with a permeability that allows the passage of water to the natural soil, reducing the volume of water to be conveyed through the sewers or diverting it quickly to conduits located under the structure, avoiding accumulations of water on the surface.

Figure N°3: Porous Asphalt Pavement (Source: Aggregate Inc http://www.aggregate.com).
Figure N° 4: Pervious concrete (Source: perviouspavement.org)

• Ponds: These are permanent artificial rainwater reservoirs, between 1 and 2m deep, with emergent and submerged vegetation. These types of ponds are designed to ensure water retention during long periods of rainfall.

Figure N° 5: Artificial pond (Source: Minnesota Department of Agriculture: http://www.mda.state.mn.us).

- Geo-cellular or modular systems are systems used to buffer or store rainwater, these systems can be soakaways or storage volumes. They are resistant to traffic loads and can be combined with permeable paving systems to optimise their performance and can be used under roads or car parks.

Figure N° 6: Rainwater buffer structure using plastic modules (Source: Hydro International: http://www.esi.info).

1.1. **Sponge City**

The "Sponge City" concept was formed in response to China's current water problems, as the country is facing problems of water scarcity and pollution, flooding and aquatic habitat degradation [9]. The Sponge City concept has three main principles, ecological water management, "green" infrastructure and permeable urban pavements [10].

Today's "sponge cities" could recycle approximately 70% of rainwater by implementing improvements in surface permeability, detention systems, storage, purification and urban water drainage systems. Some of the most commonly used technologies in such cities are green roofs, green spaces, artificial water surfaces, infiltration ponds, biological retention facilities and permeable pavements [10].

Interventions in sponge city projects include the following:

• Open and continuous green spaces: these can include interconnected watercourses, canals and ponds through neighbourhoods that can naturally detain water and filter it, as well as foster urban ecosystems, boost biodiversity and create cultural and recreational opportunities.

- Green roofs that can retain rainwater and naturally filter it before it is recycled or released into the environment.
- Design of city interventions, including the construction of sustainable drainage and bioretention systems to detain stormwater runoff and allow infiltration into groundwater, porous roads and pavements that can safely accommodate vehicular and pedestrian traffic while allowing water to be absorbed, drainage systems that direct stormwater runoff into green spaces for natural absorption
- Water saving and recycling, including grey water recycling at building block level, encouraging consumers to save water, awareness campaigns and improved smart monitoring systems to identify leaks and inefficient water use.

1.2. **Local measures**

Finally, various measures taken in the city of Santa Fe to mitigate the negative environmental, social and economic impacts of rainfall are presented.

The Rainwater Drainage Master Plan is a work developed by the National Water Institute (INA), with the collaboration of the Government of the city of Santa Fe, which was initiated in December 2007. Part of the works included in the Master Plan are specified in Ordinance N°12 194/15, in the so-called Plan de Obras Localizadas de Desagües.

Another intervention is the rainwater surplus regulation system, included in Ordinance N ° 11.959/13. The purpose of this ordinance is to establish a regulatory framework for the incorporation of rainwater surplus regulation systems, in order to contribute to the optimisation of the operation of the urban storm drainage system in the city of Santa Fe.

In its articles, it presents modifications to the soil waterproofing factors, which is the ratio between the projection, on an ideal plane at +/- 0 ground level, of the built-up covered surface plus the surface of the flooring or paving and the surface of the plot. In addition, when the waterproofing factor requirements are not met, hydraulic devices must be incorporated to regulate the evacuation of rainwater surpluses in order to reduce their impact on the urban storm drainage system. These devices must produce at least a fifty percent (50%) reduction in the maximum flow to be evacuated. Also included are obligations for owners of properties with impermeable surfaces greater than 1000 m^2 , who must propose a flow regulation system, among other measures.

In accordance with the provisions of decree D.M.M. N°00701/13, the city of Santa Fe is located in the flood valley of the Paraná River, surrounded by numerous watercourses, with the consequent periodic variations of flows and levels, which makes it a city at permanent water risk.

Therefore, by means of the Urban Development Regulation (Ordinance N°11.748), the creation of fifty-two hectares of reservoirs and the requirement to comply with a Soil Impermeability Factor (FIS) appropriate to each zoning of the city, the reduction of plastic bags and a special regime for Large Waste Generators, with the aim of helping to avoid the obstruction of rainwater drainage systems and the inclusion of Green Ribbons on the pavements (portion destined for grass and trees to improve soil absorption) were planned.

Resolution N°14.716/13 establishes that the Municipal Executive Department will proceed to formalise an agreement with the National Water Institute (INA), with the aim of carrying out studies and quantitative evaluations aimed at identifying critical points of interest located in public spaces that are susceptible to being retarded, and to establish an order of priorities for intervention, according to the percentage of waterproofing in each basin, based on what is projected in the Rainwater Drainage Master Plan, drafted in due course by the aforementioned

body, in order to minimise the effects of heavy rainfall.
As a result of this resolution we can mention the report entitled "Study of critical areas for frequent flooding in the city of Santa Fe" [10], completed in May 2015, whose study area includes the twenty-six (26) basins defined in the aforementioned plan.
The vast majority of the solutions proposed by the Master Plan include interventions related to the extension and implementation of trunk and secondary conduits of the storm drainage network, i.e. conventional infrastructure alternatives. However, projects that could be defined within the usual typologies of Sustainable Urban Drainage Systems are also proposed; these alternatives were presented by the INA in a second part of the report in July 2017 [11]. Among the latter are the following:
1.1.1. <u>Cuenca Unión: in order</u> to mitigate the effects of heavy rainfall in this area, alternatives were analysed that contemplate regulating the inflows by means of a reservoir to be built in Plaza Constituyentes along the perimeter pavement (Figure N°7). **(Projected intervention).**

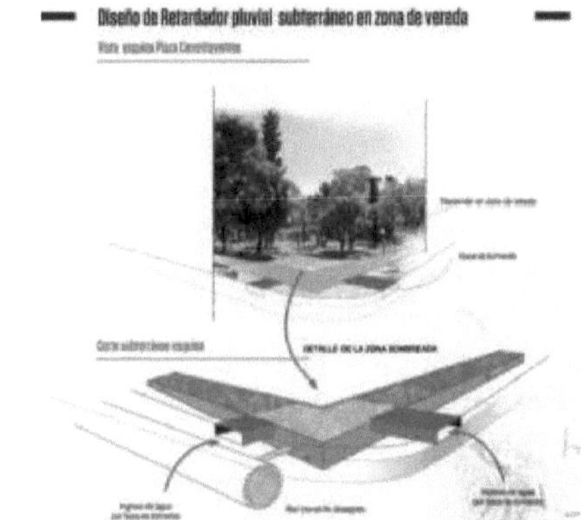

Figure N°7: Pluvial retarder project Plaza Constituyentes. City of Santa Fe [12].

Figure N°8: Location plan of the Union basin [13].

1.1.2. <u>Durán - Parque basin:</u> for this basin, the construction of the secondary basin on Calle Catamarca (Master Plan) was considered, in addition to the construction of an underground reservoir in the central flower bed of Avenida Freyre with a rectangular section and a total volume of 3000 m^3 (Figures N°4 and 5). **(Intervention carried out).**

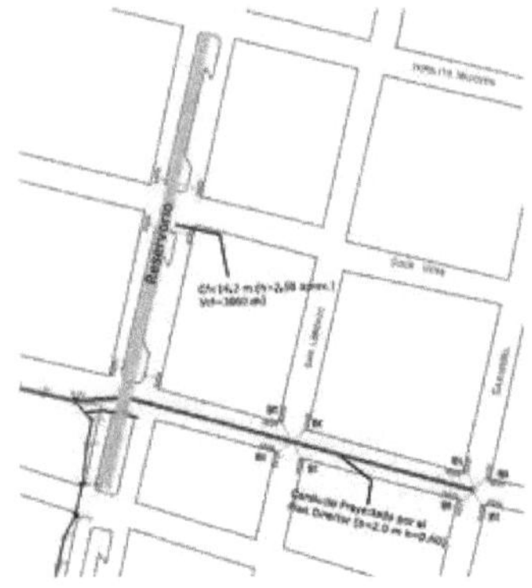

Figura N°9: Esquema y ubicación del reservorio [12].

Figura N°10: Reservorio construido en funcionamiento. Ciudad de Santa Fe [14].

Figure N°9: Reservoir schematic and location [12].

Figure N°10: Constructed reservoir in operation. City of Santa Fe [14].

Figure N° 11: Location plan of the Duran - Parque basin [13].

1.1.3. Cruz Roja basin: in this basin it is proposed to regulate the inflow flows by means of a reservoir in Plaza del Soldado, which has a large space such as an old underground construction that constitutes a very important relative advantage compared to other regulation sectors proposed due to the possibility of implementing an underground regulation work consisting of a regular shaped basin of considerable dimensions that can be directly connected to an existing conduit on the north pavement of Salta street. **(Projected intervention).**

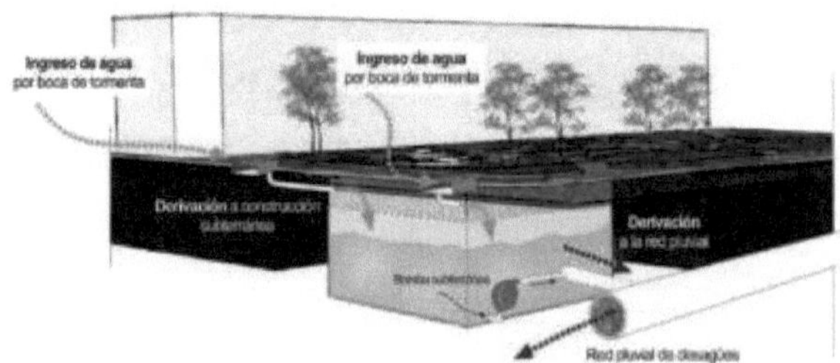

Figure N°12: Reservoir schematic [12].

Figure N° 13: Location plan of the Red Cross basin [13].

Chapter II

Logical framework approach.

The Logical Framework methodology was developed in 1969 by the United States Agency for International Development. It is currently used by various multilateral organisations such as the Inter-American Development Bank, the Spanish Agency for International Development Cooperation, among others.

This methodology can be described as a tool that facilitates the process of project conceptualisation, design, implementation and evaluation, emphasising goal orientation, target group orientation and facilitating participation and communication among stakeholders. This analytical tool is used to improve project planning and management, based on the following tasks:

- Clarify the purpose and justification of a project.
- Identify information needs.
- Clearly define the key elements of a project.
- Analyse the project environment from the outset.
- Facilitate communication between the parties involved.
- Identify the key variables to measure the success or failure of a project.

It also provides multiple advantages over less structured approaches, such as:

- It provides a uniform terminology that facilitates communication and serves to reduce ambiguities.
- It provides a format for reaching precise agreements on project objectives, goals and risks that are shared by the different actors involved in the project.
- It focuses the technical work on the critical aspects and can shorten project documents considerably.
- It provides a structure to express, in a single table, the most important information about a project.

The main parts of the LFA are the identification of the problem, the analysis of stakeholders, objectives and alternatives, the identification of the main elements of the project through the Planning Matrix and the identification of the External Factors.

Problem identification consists of establishing cause-effect relationships between the negative factors of an existing situation. Here we seek to concretely define the Problem Situation (PS), Problem Need (PN) or Opportunity for Improvement (OP). The problem should be defined as a gap or deficit and presented as a real negative state located in a target population. The solution should not be incorporated into the problem statement, but only the focal or central problem should be stated.

In order to do this, a set of questions must be answered in order to identify the main problem. What negative consequences/effects or manifestations are perceived by a given social group? Who and/or what things are affected by these negative consequences? What are the singularities of the negative consequences? Are they permanent, random, periodic? Where do the negative consequences occur? What are the impacts of these negative consequences? What are the impacts? Their succinct description. Prognosis of the evolution of the situation without intervention.

Likewise, a problem is not the absence of a solution, but an existing negative state. In the analysis it is useful to differentiate between the causes of the problem, the problem itself and its effects or consequences. The so-called "problem tree" can be used to organise the ideas. It is intended to locate the central or focal problem to be solved by the project. The problem tree has three different levels: the causes of the problem, the problem itself and its effects or consequences.

The stakeholder analysis seeks to identify all groups, organisations, people related to and affected by the situation/problem under consideration. Existing or potential conflicts between them are raised and investigated. The interests of these groups in relation to the identified problem and their perceptions of the problems are also investigated. The available resources (political, legal, human, financial, etc.) are observed. Finally, a problem tree is drawn up.

After having identified the main problem, the analysis of objectives is carried out, where possible solutions are proposed. Based on the problem tree, the interventions to be carried out to solve or reduce the main problem are identified, so that the negative states of the diagram become positive states achieved as a result of the intervention carried out in the system.

Once the objectives have been determined, an analysis of the alternatives is carried out, which consists of setting out the possible solutions that could achieve the proposed objectives, the capacity of the organisation that is going to develop the project, the means at its disposal, the resources that it can reasonably handle and the environment surrounding the project, within this analysis there is the possibility of eliminating those objectives that cannot be achieved under the conditions assumed for the project (lack of executive capacity, financing, legislation, available technology, etc.).

Any project must reflect the possible alternatives and justify the choice of one of them on the basis of criteria used to compare each of the options. This analysis can be very complex depending on the criteria used at the time of comparison. The option adopted must be feasible, understood as the possibility of implementing the selected alternative and presenting the best values for the comparison variables. A matrix of the different options and the chosen criteria can facilitate such an analysis.

1. MLE development

The Logical Framework Approach (LFA) is applied to this project in order to improve the planning and efficiency of the project.

1.1. Problem definition

Increased rainfall intensities, as a consequence of climate change, produce higher peak flows than those projected at the time of urban storm drain design. This is compounded by increased surface imperviousness, which leads to changes in surface runoff volumes, runoff velocities and direct infiltration rates.

This causes the collapse of the drainage system, which is forced to evacuate flows in excess of the design values. Santa Fe and most localities in the province suffer from this phenomenon on a regular basis. In addition, the problem is aggravated by the increase of waste in storm water, caused by the increase in population and population density in the cities, which generates obstructions and damage to the storm water infrastructure.

The latter also has a significant environmental impact, as waste washed away by rainwater is deposited in watercourses, decreasing water quality, damaging the ecosystem, causing health risks, etc.

What consequences/effects or negative manifestations are perceived by a given social group?

- Pollution in the Setúbal Lagoon and riparian areas.
- Localised flooding.
- Inadequacy of planned infrastructure capacity.
- Waterlogging of streets and pavements.

- Foul odours from sewers.
- Material and economic losses.
- Affects on traffic and productive activities.

Who and/or what are affected by these negative consequences?

Although the problem to be addressed is of a general or global nature, the particular situation of the city of Santa Fe, which has approximately 391,164 inhabitants, is analysed [15].

The main people affected are those living in the critical areas identified by the National Water Institute (INA).

What are the singularities of negative consequences: are they permanent, random, periodic?

- Pollution in the Setúbal Lagoon and riverside areas. **PERMANENT.**
- Localised flooding. **JOURNAL.**
- Inadequacy of planned infrastructure capacity. **PERIODIC.**
- Waterlogging of streets and pavements. **JOURNAL.**
- Bad smells coming from sewers. **JOURNAL.**
- Material and economic losses. **JOURNAL.**
- Traffic and productive activities affected. **JOURNAL.**

Where do the negative consequences occur/occur?

- Setúbal Lagoon.
- Streets, avenues, pavements, squares and other public spaces.
- Private properties.
- City of Santa Fe.
- Critical areas determined by the National Water Institute (INA).

What do these negative consequences impact on?

The negative consequences identified have a fundamental impact on the daily life of citizens, generating minor inconveniences with a high frequency and serious affectations periodically, producing economic affectations such as material losses, major flooding for prolonged periods of time or even loss of life. The future expectation of possible inconveniences caused by rainfall affects investment in the city, the value of land and also has a negative impact on the psyche of citizens, generating situations of anxiety and/or panic that negatively affect the quality of life and economic activities.

What are the impacts?

- Pollution in the Setubal Lagoon and riverside areas: waste that is discarded on the public highway is washed into the drains and ends up in storm drains, reservoirs or against the grates of pumping stations, and is then discharged into natural watercourses such as the Setubal and the Salado River.
- Localised flooding: These phenomena are a consequence of the situation previously mentioned, which have an impact on the normal functioning of the city by causing flooding and the effects that this entails (street closures, suspension of bus lines, impassable unpaved streets, etc.).
- Inadequacy of the projected infrastructure capacity: given the increase in the rate of waterproofing of the surface, there is an increase in the peak flows to be delivered, as the absorption of the natural soil decreases. Added to this is the variation in intensities caused by the phenomenon of climate change. Both situations create a scenario in which the conduction capacity of the projected infrastructure is greatly exceeded.
- Waterlogging in streets and pavements: the presence of urban waste in the drainage ducts causes total or partial obstructions, which, together with the increase in the flow of water to be

discharged, causes overflows.

- Material and economic losses: these overflows cause traffic disruptions, floods that cause damage to public and private property and/or loss of life, resulting in considerable economic losses for the community.
- Bad odours from sewers: urban waste contains a large amount of organic waste [16], which, as a result of blockages, can be retained in the drains for long enough to decompose, and current structures allow animals such as rodents to enter the sewage system, where they feed, reproduce and eliminate their waste inside the pipes, exacerbating the problem.

1.2. Prognosis for the evolution of the situation without intervention

There are a number of factors that cannot be addressed at the local, regional or national level, which affect the entire world population, such as climate change or population growth. These situations lead to an increase in urban waste and waterproofing rates, which can be comprehensively addressed with future planning, awareness campaigns and regulations to improve the existing situation.

If measures are not taken to reduce both urban waste and surface waterproofing, the associated problems described above will continue to increase until the quality of life of the population worsens considerably, producing huge economic and environmental losses. Once this situation is reached, the measures to be taken to prevent further damage and improve the existing state will have to be structural and extremely costly.

1.3. Table of MLE participants

GROUP	RELATIONSHIP WITH THE PROJECT	INTERESTS AND NEEDS	PERCEIVED PROBLEMS	PROJECT INPUTS OR CONSTRAINTS
City dwellers	Indirect beneficiary	Cesar to stop suffering damage and inconvenience due to storm water overflows	• Pollution in the Setúbal Lagoon and riparian areas. • Localised flooding. • Foul odours from sewers. • Material and economic losses.	Unidentified
Public agencies responsible for the design, construction and maintenance of storm drains.	Direct beneficiary	Minimise the costs of execution, maintenance, refurbishment, etc. of works.	Obstructions and deterioration of ducts	Lack of knowledge for the application of a non-traditional system
Private companies involved in storm drainage construction	Direct beneficiary	Making money by carrying out works	Unidentified	Unidentified
Neighbours of the construction site	Direct beneficiary	Cessation of damage or inconvenience from stormwater overflows	• Localised flooding. • Foul odours from sewers. • Material and economic losses. • Missing fences due to vandalism.	Adversely affected in the construction period
Concrete processing plants	Direct beneficiary	Increase concrete sales. Commercialisation of permeable concrete	Unidentified	It provides the supply of permeable concrete. As it is not a widely used material, it is difficult to guarantee the quality of the product.
Municipality of the city of Santa Fe	Direct beneficiary	Guaranteeing compliance with	Increase in number and frequency of complaints from	Application of drainage concrete in existing and

		current regulations. Seeking to improve the quality of life of citizens.	residents about drainage problems.	future infrastructure
GROUP	**RELATIONSHIP WITH THE PROJECT**	**INTERESTS AND NEEDS**	**PERCEIVED PROBLEMS**	**PROJECT INPUTS OR CONSTRAINTS**
Fishermen	Indirect beneficiary	To obtain fish in sufficient quantity and quality to make their activity profitable.	Decrease in the amount of fish present in the Setúbal Lagoon	Unidentified
Entities responsible for providing drinking water	Direct beneficiary	Lowering the costs of water purification	Increased levels of contamination of incoming water to the plant	Encourage the use of draining concrete components, given their filtering capacity.

Table N° 1: Table of participants in the logical framework approach, summarising the interaction between each group analysed and the project.

1.4. Assessment of importance and influence

The following diagram represents the groups of influence, those who will be affected to a greater or lesser extent by the project, whether negatively, neutrally or positively, differentiated by colour and organised according to the magnitude of the impact and influence generated.

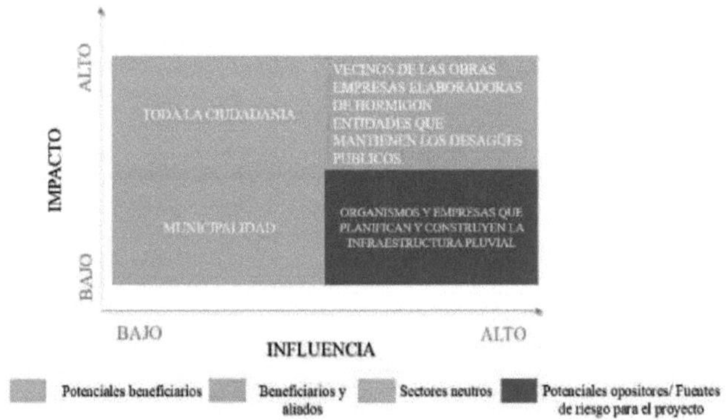

Figura N° 14: Esquema grafico donde se representa la importancia e influencia percibida que la ejecución del proyecto va a tener en los involucrados.

1.5. Identification of risks and assumptions

GROUPS INVOLVED	RELATIONSHIP WITH THE PROJECT	CAUSES	CONTRIBUTIONS	RESTRICTION
City dwellers	Potential beneficiaries	Improving quality of life	Unidentified.	Unidentified
Private designers and builders of public works	Direct beneficiaries	New alternative to reduce a problem	Adoption and dissemination of the proposed alternative	Possible lack of interest in the implementation of a non-traditional system
Responsible for the design and construction of stormwater facilities	Potential opponents	Adaptation to a new system	Unidentified	Possible lack of interest in the application of a non-traditional system

				and/or non-adaptation to it Visualisation of the system as a competence in the field of work
Neighbours using the works	Direct beneficiaries	Variety of alternatives for the same problem Improved infrastructure	Financial resources	Lack of interest in the application of a non-traditional system in case the costs are higher than traditional solutions. Possible dissatisfaction with the inconveniences of the construction phase
Concrete batching plants	Direct beneficiaries	New demand for its products	Incorporation of drainage concrete in its mixes range	Training and/or counselling needs
Municipality	Neutral sector	Unidentified	Unidentified	Need to ensure approval/inspection of modules to enable system adoption

Table N° 2: Summary of the relationship between the different groups involved in the project, their contributions and/or constraints to the project.

1.6. Problem tree

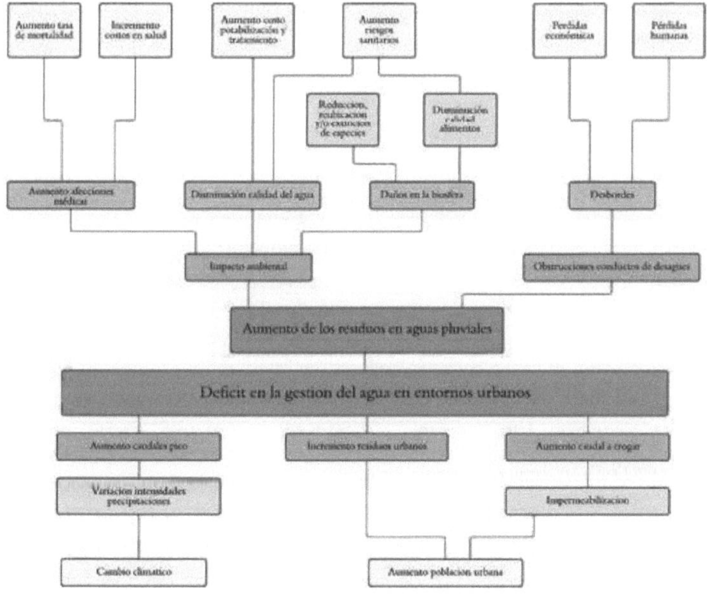

Figure N° 15: Problem tree

1.7. Objective tree

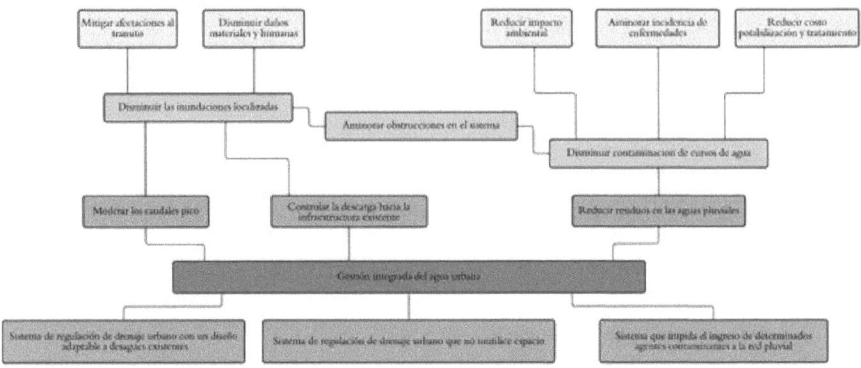

Figure N° 16: Objective tree.

2. *Analysis of alternatives*

Given that the situation analysed presents a wide variety of problems, which cannot be solved in their entirety with a single intervention, an analysis is carried out to evaluate different alternatives and to obtain the most efficient one according to the objectives pursued.

To do this, a multi-criteria matrix is used, designed to quantify and weight different aspects based on the problems posed and their relationship with each alternative. This results in a hierarchical positioning of the alternatives, which will indicate the order of action in which the causes should be addressed to provide an efficient solution to the overall problem [17].

The criteria used for the weighting of each item, the percentages of incidence with respect to the total and the way in which the scoring is structured are explained.

Environmental and health impact: this criterion considers the positive or negative effects produced by each alternative on biodiversity, public health, soil and/or water pollution, etc.

Score 5: High influence.

Score 1: Low influence.

Decrease in the probability of overflows: this item analyses whether the intervention to be carried out has any effect on the capacity to deliver rainwater. A score of 1 to 5 will be given.

Score 5: High influence.

Score 1: Low influence.

Works to be carried out: the magnitude of works expected to be carried out in order to achieve a comprehensive solution to the cause represents the interventions to be carried out in the city, demolitions, work times and costs associated with the initial investment, maintenance, etc. It will be given a score from 1 to 5.

Score 5: low magnitude of work.

Score 1: High magnitude of work.

The alternatives to the problem are diverse and most of them have been described previously. Below we analyse different possible interventions that address the problem from different areas, from awareness campaigns that do not require any construction to drainage works that must intervene in a large part of the city.

Criteria Alternatives X.	Environmental impact and health [50%].	Decrease in the probability of overflows [30%].	Work to be carried out [20%].	Result

Reduction of waste and pollutants in stormwater with drainage concrete	5	2	5	4.1
Expansion of stormwater infrastructure	1	5	1	2.2
Increased absorbent surface	1	3	3	2
Construction of reservoirs	2	4	4	3
Awareness-raising campaigns on urban waste, responsible use of water, etc.	3	1	1	2

Table N° 3: Multi-criteria matrix.

Chapter III

Project definition and explanation

1. Selection of alternative

The alternative chosen and developed in this project is based on the application of drainage concrete, with the aim of being used as a retention device for urban waste at the entrances to sewers and for regulating rainwater surpluses, contributing to the existing regulators in accordance with Decree D.M.M. N°00701/13 of the Municipality of Santa Fe.

When analysing its performance in reducing the environmental impact of the current problematic situation, it is perceived that it contributes considerably to reducing the pollution of the water discharged, as it prevents most urban waste from passing into the public sewage system, preventing it from being discharged together with rainwater into the receiving watercourses, in addition, surface runoff generally contains suspended solids, organic compounds and heavy metals, which come from vehicles, the surrounding environment and the pavement itself, permeable pavements purify water by mechanical retention, physical adsorption, chemical reaction and biodegradation [18].

With regard to its influence on reducing the probability of future overflows, it is considered that by reducing the entry of objects into the conduits of the storm drainage system, the obstructions produced by them are reduced, This increases the section available for water to be channelled out of the city and, with the water retained in its pores, delays the entry into the system of a percentage of the volume of precipitation, and when the weather is dry, the pavement materials can evaporate the water to the outside.), structural characteristics (porosity), water content, etc. [18].

It can be seen that the works to be carried out are on a small scale, as they consist of replacing the existing metal gratings with drainage concrete modules. This requires minimal intervention to the existing infrastructure and does not require a great deal of time, machinery or manpower.

2. Location and current situation of the intervention area

The sub-basin chosen for the intervention is represented in the following figure (figure N° 10), it is located within the Plaza España basin, which has an area of 161 hectares, located in the centre of the city of Santa Fe [13].

From the map provided by the Secretariat of Water Resources of the Municipality of Santa Fe, the sub-basin is located on San Martin street, between Lisandro de la Torre and Tucumán streets approximately, in the area where only pedestrian traffic is allowed. The chosen sub-basin has an area of 3.81 hectares and the waterproofing factor of the soil is 100%, so that all the precipitated water will be transformed into surface runoff. The length of the project is approximately 360 metres.

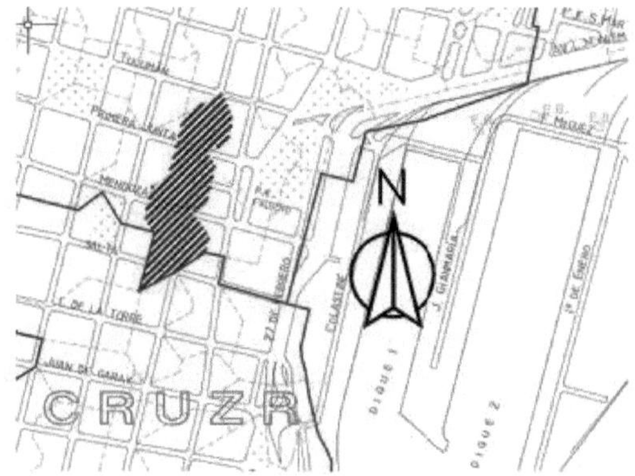

Figure N° 17: Diagram of the sub-basin to be intervened, general drainage plan (Municipality of Santa Fe) [13].

Currently, the excess rainwater from Santa Fe's pedestrian street enters the storm drainage system through small surface channels, covered by metal grates with openings 20 centimetres wide by 2 centimetres long, which allow water and a large amount of waste to pass through to the sewers. There is also the phenomenon of vandalism and theft of metal parts of the public sewers, a problem that would be reduced if they were replaced by concrete modules [19].

Figure N° 18: View of drainage gratings, Santa Fe pedestrian walkway.
Figure N° 19: View of drainage gratings, Santa Fe pedestrian walkway.

3. *General objectives and scope of the project*

The project is presented on the basis of various benefits that would favour the population involved, and it is also noted that the knowledge of the material (drainage concrete) and the techniques to be used for its production and application are sufficiently developed in the city of Santa Fe to be able to carry out the project.

The main objective of the presence of concrete drainage elements in the area where excess rainwater enters the drainage network is that they reduce the entry of urban waste into the network, preventing it from being deposited in natural watercourses, improving the quality of

the water discharged and reducing obstructions and maintenance tasks in the drainage network.
On the other hand, the aim is also to increase the durability of the drainage structure by proposing a technically and economically viable alternative to replace the steel gratings that must be protected with anti-rust paint periodically and are prone to being stolen and/or vandalised, while the projected modules require less maintenance and due to their material (price), their reduced possibility of use in another place, weight and dimensions reduce the probability of theft.
A third purpose of the application of drainage concrete modules is to contribute to the reduction of odours and noise produced by sewers, thus improving the comfort of pedestrians. It greatly reduces the reproduction of animals, insects and other living beings in the vicinity of the surface, helping to reduce the likelihood of diseases or infections caused by them [16]. Likewise, the use of the projected modules provides uniformity of colour and materiality in the pedestrian surface, not being perceived as a drainage work and also avoiding the visual inside the sewer. It also provides pedestrian safety, as it eliminates the empty spaces in the gratings, where objects can be dropped or accidents can occur.
Another motivation is to reduce the volume of water discharged, based on the fact that the volume of voids in the draining concrete allows water to be stored, acting as a flow regulator, thus collaborating with the already collapsed storm drainage network of the city of Santa Fe. This volume of retained water will not only be subsequently released into the conduits, reducing peak flows, but part of the water will also evaporate before entering the drains, since another part of the water that reaches the draining concrete, approximately 5 to 6%, will be retained prior to evaporation as water absorption in the material [16]. There is a precedent of a Civil Engineering Degree Final Project by Aguirre Diego and Argento Romina, entitled "Drainage concrete as a regulator of rainwater surpluses" in 2021, in which the possibility of using this material as a flow regulator was studied in compliance with the corresponding municipal ordinance.
In short, the proposed alternative consists of a concrete drainage layer designed to be placed at the entrances to the urban storm drainage system, capable of preventing the passage of waste into the conduits, thus avoiding obstructions in them and the environmental impact caused by waste being deposited in the watercourses, among other objectives. Likewise, due to its porous structure, it allows rainwater to be stored inside and gradually evacuated towards the drains, thus regulating the peak flows produced by rainfall. The modules will be designed to be applied at the entrances to the watercourses, among other objectives.

existing storm drains in the pedestrian walkway of the city of Santa Fe, allowing light vehicles to pass over them.

The geometric design of the module, the hydraulic analysis to verify its capacity to deliver the necessary flow for the design storm, the mechanical analysis to check its resistance to the planned stresses, filtration capacity and the calculation of the materials necessary for the proposed intervention, together with an estimate of construction times, necessary investment and comparison of costs per square metre with the structures available on the market, the influence of clogging, durability, among other aspects concerning the project, will be carried out.

Chapter IV

Drainage concrete as a material

1. Historical overview.

The initial use of pervious concrete was in the UK in 1852, with the construction of two residential houses and a sea wall. Cost efficiency seems to be the main reason for its first use, given the limited amount of cement used. This material consisted of only coarse gravel and cement and was not mentioned in the published literature until 1923, when a group of 50 two-storey houses were built in Edinburgh, Scotland and pervious concrete re-emerged as a viable building material. By 1942 it had been used to build over 900 houses [20].

Its use grew steadily in Europe, especially during World War II, as pervious concrete uses less cement than conventional concrete and cement was in short supply at the time, coupled with the vast housing needs, which encouraged the development of new methods, or methods that had not previously been used in building construction. With pervious concrete, less cement was used per unit volume of concrete compared to conventional concrete, and the material was advantageous where labour was scarce or expensive.

In Germany, after the war, this system was used because the disposal of large quantities of brick rubble was a problem, leading to research into the properties of pervious concrete. Elsewhere, the unprecedented demand for brick, and the subsequent inability of the industry to ensure an adequate supply, led to the adoption of pervious concrete as a building material. It then continued to gain popularity and its use spread to areas such as Venezuela, West Africa, Australia, Russia and the Middle East, the first report on the use of pervious concrete in Australia being in early 1946. After World War II it became widespread for applications such as hollow load-bearing walls, precast panels and blocks, load-bearing walls for buildings up to 10 storeys and as infill panels in high-rise buildings [21].

It was also widely used for industrial, public and domestic buildings in areas north of the Arctic Circle because the use of traditional building materials proved impractical. Examples of these drawbacks include the high transportation costs of brick, the fire hazards of wood, and the poor thermal insulating properties of plain concrete.

Although not a new technology, pervious concrete has received renewed interest in the United States, in part because of Federal Clean Water Legislation. The US Environmental Protection Agency's (EPA) Phase II Final Rule requires operators of all municipalities in urban areas to develop, implement and enforce a programme to reduce pollutants in stormwater runoff from new development and redevelopment projects affecting an area equal to or greater than one acre (approximately 0.4 [hectares]).

This is a requirement for obtaining a National Pollutant Discharge Elimination System (NPDES) permit. Among other stipulations, municipalities must develop and implement strategies that include a combination of structural and/or non-structural best management practices (BMPs). Pervious concrete pavement is recognised as an infiltration structure BMP by the EPA as it provides the control

pollution and stormwater management at the first discharge. In addition to federal regulations, there has been a strong movement in the United States towards sustainable development, which can be defined as development that meets the needs of the present generation without compromising the needs of future generations [20].

2. Definition, main properties and uses.

2.1. Definition.

Pervious concrete is a high porosity concrete whose mix consists of coarse aggregate, cement, water, admixtures and little or no fine aggregate. The aggregates are completely covered by cement paste and are bonded to each other. The amount of cement mortar is, however, so low that the spaces between the aggregates are not completely filled, preserving the interconnectivity of the pores, allowing the flow of water or other liquids through the structure.

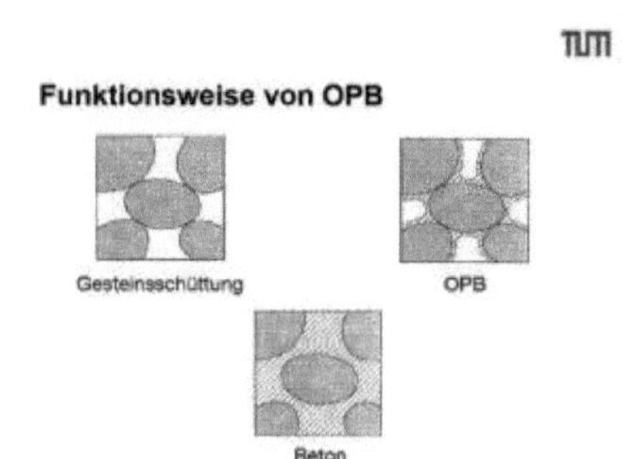

Figure N° 20: Comparison of the internal structure of loose aggregates, aerated concrete and plain concrete [22].

Figure 21 shows that by adding a layer of cement paste to the loose aggregates, the aggregates are bound together in such a way that the structure is fixed and the pores remain interconnected, differing from ordinary concrete in that the latter completely fills the voids with cementitious paste.

2.2. Main properties.

The fundamental characteristics sought in pervious concrete that modify its performance for the different applications where they are used are the size and type of stone, the water-cement ratio, void content, compressive strength, flexural strength, etc., modulus of elasticity, density and infiltration rate, among others, such as average pore size, consistency, etc.

These data can be extracted from the following table and will be taken into account when designing the reception structure concerning the project.

Offenporiger Beton: Lärmmindernde Fahrbahndecken

Quelle: heidelbergcement

Festigkeitsklasse	C20/25 - C25/30	Richtwerte	Prüfergebnisse
Gesteinskörnung	gGK 5/8 mm, WS (Splitt) fGK 0/2 mm (Quarzsand)	1450 - 1600 kg/m³ 50 - 70 kg/m³	1464 kg/m³ 68 kg/m³
Zement	CEM I 32,5 R / -42,5 N oder R	300 - 350 kg/m³	350 kg/m³
Wasser	(Frischwasser)	35 - 40 kg/m³	40 kg/m³
Polymerdispersion	18 - 20 M.-% v. Z.	60 - 70 kg/m³	62 kg/m³
Zusatzmittel	FM (ggf. VZ)	1 - 3 kg/m³	1 kg/m³
w/z-Wert		0,27 - 0,30	0,29
Fasern	PAN, PVA, AR-Glas (6-12 mm)	2 kg/m³	2 kg/m³
Konsistenz	v	1,30 - 1,34 (C1)	1,29 (C1)
Hohlraumgehalt	(Tauchw. DIN EN 12390-7) P	18 ± 3 Vol.-%	15 Vol.-%
Druckfestigkeit	$f_{ck,cube}$	≥ 25 MPa	32,4 MPa
Biegezugfestigkeit	$f_{ct,bz}$	≥ 4,5 MPa	5,1 MPa
Spaltzugfestigkeit	$f_{ct,sz}$	≥ 2,7 MPa	3,4 MPa
statischer E-Modul	ε_B	20000 - 24000 MPa	20900 MPa
Rohdichte Festbeton	ρ_B	1,90 - 2,15 kg/dm³	2,07 kg/dm³

Figure N° 21: Table of values associated with porous concretes. Chair Special Concrete. Technical University of Munich 2019. [22]

Hormigón poroso: Pavimentos con disminución de ruidos

Quelle: heidelbergcement

Tipo de resistencia	C20/25 - C25/30	Valor de referencia	Resultados
Tamaño de agregado	gGK 5/8 mm, WS (Splitt) fGK 0/2 mm (Quarzsand)	1450 - 1600 kg/m³ 50 - 70 kg/m³	1464 kg/m³ 68 kg/m³
Tipo de cemento	CEM I 32,5 R / -42,5 N oder R	300 - 350 kg/m³	350 kg/m³
Agua	(Frischwasser)	35 - 40 kg/m³	40 kg/m³
Polimero de disperción	18 - 20 M.-% v. Z.	60 - 70 kg/m³	62 kg/m³
Aditivo	FM (ggf. VZ)	1 - 3 kg/m³	1 kg/m³
Relación agua cemento		0,27 - 0,30	0,29
Fibras	PAN, PVA, AR-Glas (6-12 mm)	2 kg/m³	2 kg/m³
Consistencia	v	1,30 - 1,34 (C1)	1,29 (C1)
Contenido de vacios	(Tauchw. DIN EN 12390-7) P	18 ± 3 Vol.-%	15 Vol.-%
f'c	$f_{ck,cube}$	≥ 25 MPa	32,4 MPa
Resistencia a flexión	$f_{ct,bz}$	≥ 4,5 MPa	5,1 MPa
Resistencia a corte	$f_{ct,sz}$	≥ 2,7 MPa	3,4 MPa
Módulo de Young	ε_B	20000 - 24000 MPa	20900 MPa
Densidad	ρ_B	1,90 - 2,15 kg/dm³	2,07 kg/dm³

Figure N° 21.1: Translated table of values associated with porous concretes. Chair Special Concrete. Technical University of Munich 2019. [22]

From the above tables, the modulus of elasticity, bending strength and shear strength data will be extracted as reference, which will be taken into account at the moment of designing the reception structure concerning the project and evaluating it mechanically.

<u>Modulus of elasticity:</u> 20000 - 24000 [Mpa].

<u>Water-cement ratio:</u> 0.27 - 0.3.

Figure N°22 shows the relationship between the water/cement ratio and void content of a draining concrete mix (with constant cement and aggregate contents) at two different compaction levels. Experience has shown that w/c ratios between 0.26 to 0.45 provide good aggregate cover and good paste stability.

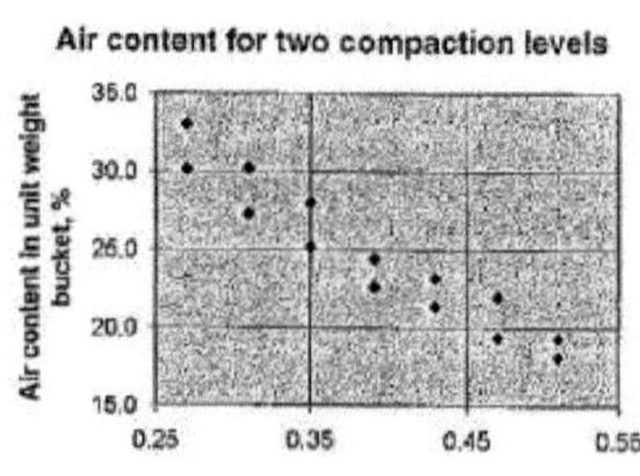

Weter-Cement ratio (w' c)

Figure N° 22: Relationship between water/cement ratio and void content [23].

Void ratio: between 18 ± 3 Vol. -%.

Density: $1,90\text{-}2,15\left[\frac{kg}{dm^3}\right]$.

The density of fresh draining concrete can be determined by ASTMC1688/ C1688M "Standard Test Method for Density and Void Content of Freshly Mixed Pervious Concrete", and is directly related to the void content of a given mix.

There are two additional methods that determine the porosity of hardened concrete. The first method involves a volumetric procedure in which the mass of water filling a sealed draining concrete sample is converted into an equivalent volume of pores. In the second method, an image analysis procedure is used on samples that have been impregnated with a low viscosity epoxy.

The accessible porosity is a function of the size of the aggregates and the relative amounts of the different sizes in the mix.

Void content is highly dependent on several factors: aggregate gradation, cementitious material content, water/cement ratio, and compaction effort.

The influence of aggregate gradation on porosity for laboratory prepared samples is shown in Figure N°23.

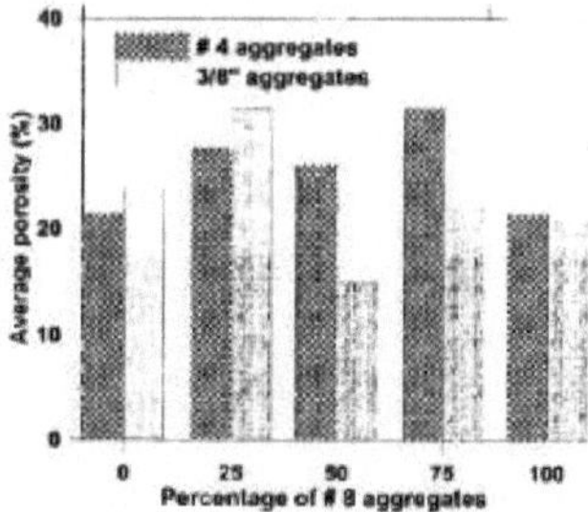

Figure N° 23: Influence of aggregate gradation on porosity [23].

It has been observed that concrete with a smaller aggregate size and a larger amount of voids has a better purification effect on air and water due to the increase in the specific surface area of the material. On the other hand, since infiltration and retention are contradictory concepts, it is necessary to co-ordinate the relationship between them, it has been found that draining concretes with smaller aggregate size could efficiently relate the two concepts.

infiltration and retention values, allowing water to pass through the structure but with a good retention rate [16].

Compressive strength: > 25 [Mpa].

The compressive strength of this type of concrete is strongly affected by the mix ratio and the compaction effort during placement.

Figure N° 24 shows the relationship between compressive strength and void content.

The graph originates from a series of laboratory tests for which two sizes of coarse aggregate were used and the compaction forces and aggregate gradation were varied. Figure 25 illustrates the relationship between compressive strength and unit weight.

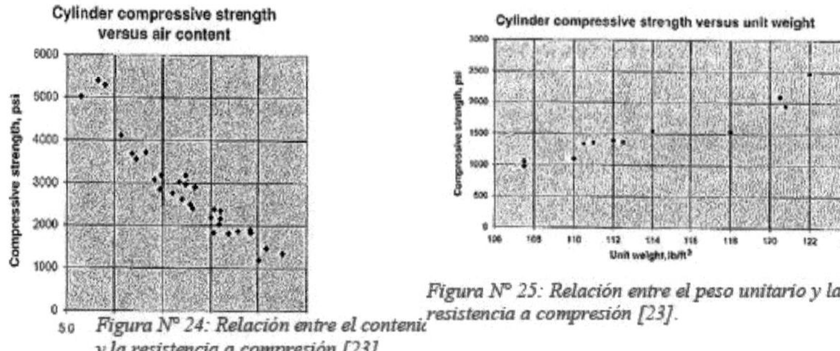

Figure N° 24: Relationship between void content and compressive strength [23].
Figure N° 25: Relationship between unit weight and compressive strength [23].

Bending strength: > 4.5 [Mpa].

Figure N° 26 shows the relationship between the flexural strength of the draining concrete and the void content based on the beam samples analysed in the same series of laboratory tests described for Figure N° 24.

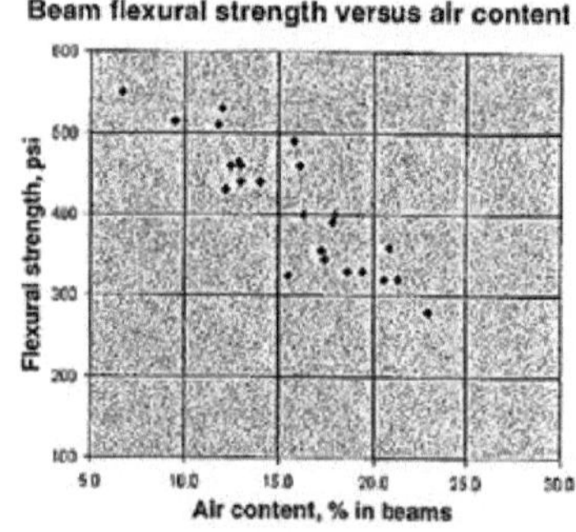

Figure N° 26: Relationship between void content and flexural strength [23].

Pore sizes:

To generate larger pore sizes in the material, larger aggregate sizes are recommended. Figures N° 27 and N° 28 represent the influence on the pore sizes of concrete of mixing two different aggregate sizes in varying proportions and one aggregate size respectively.

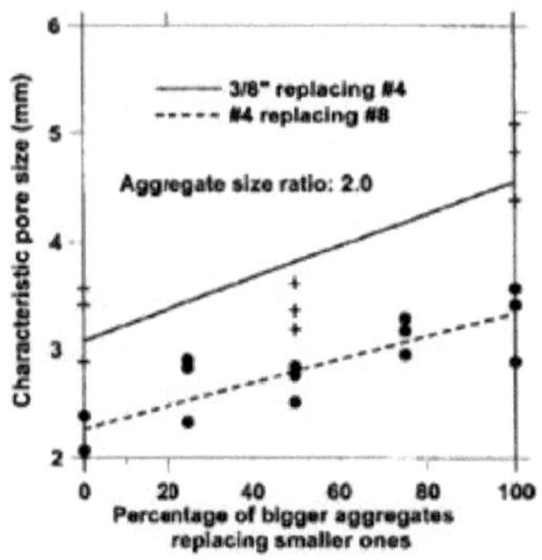

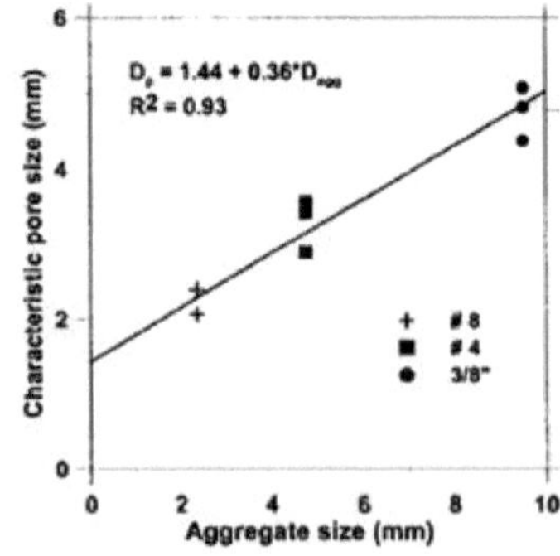

Figure N° 27: Influence of aggregate size on pore size [23].
Figure N° 28: Influence of aggregate size on pore size [23].

Replacing the smaller aggregates with an increasing percentage of larger aggregates increases the pore size, because the coarser particle introduced may not fit into the void left by the finer particle removed.

Infiltration rate: between 80 and 720 $\left[\frac{L}{\min x\, m^2}\right]$ [24]. [24].

One of the most important characteristics of draining concrete is its ability to filter water through the matrix. The infiltration rate is directly related to the void content, Figure N° 29 shows the relationship between void content and infiltration rate. Since this rate increases as void content increases, and consequently compressive strength decreases, the challenge of mix design is to achieve a balance between acceptable percolation rate and compressive strength.

In addition to porosity and pore size, a crucial factor influencing permeability is the degree of pore network connectivity. There is no simple methodology for measuring pore connectivity but it is anticipated that the use of techniques such as X-ray computed tomography will lead to an accurate determination of this parameter.

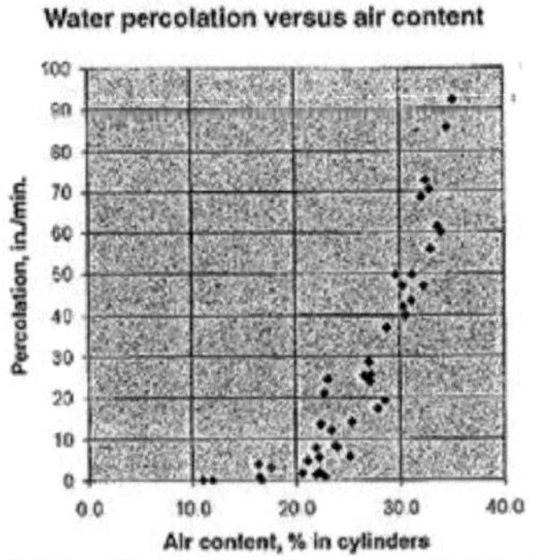

Figure N° 29: Relationship between void content and infiltration rate [23].

Sound absorption: Due to the presence of a large volume of interconnected pores of considerable size in the material, drainage concrete is highly effective in sound absorption. The pores absorb sound through internal friction between moving air molecules and their walls. The variables that influence the degree of sound absorption are void content, tortuosity and layer thickness. The following graph shows the degree of absorption α for a draining concrete sample and different frequencies.

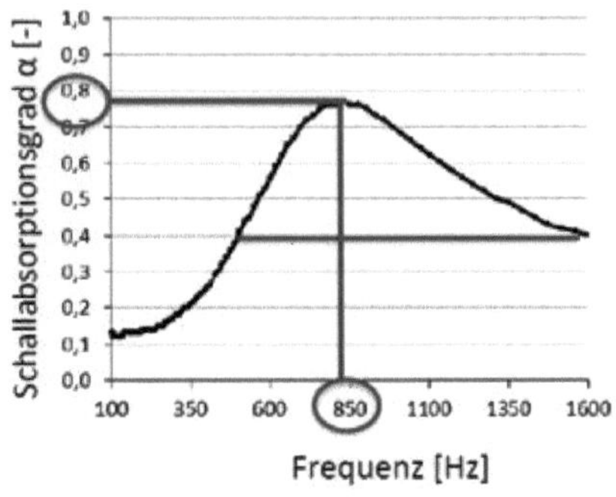

Frequency [Hz]

Figure N° 30: Relationship between the degree of sound absorption of draining concrete at different sound frequencies [22].

Each of these parameters will be decisive when designing a mix depending on the application of the material in each project. To ensure that the mixes will have a given performance, materials that meet the requirements set by the standards must be used and the dosage methods approved by the standards must be followed. The parameters for the materials used and the batching methods are developed in the following items.

2.3. Materials.

The following detailed requirements for each of the components used for the draining concrete mix are based on ACI 522R-10 (2011) [23].

Aggregates: Typically coarse aggregates are one size or graded between 3/4" and 3/8" (19 and 9.5 mm) and must meet the requirements of ASTM D448 "Standard Classification for Sizes of Aggregate for Road and Bridge Construction" and ASTM C33/C33M "Standard Specification for Concrete Aggregates".

Fine aggregates are not commonly used in draining concrete mixes because they tend to compromise the connection of the pore system. Their addition can increase strength and density, but consequently reduces the rate of water infiltration through the concrete mass. The quality of the aggregate is just as important as in conventional concrete, and needle-shaped particles and needles should be avoided. The aggregate should be hard and free of coatings, such as dust or clay or other absorbed chemicals that may adversely affect the paste or cement hydration.

The unit weight of aggregate shall be determined in accordance with ASTM C29/C29M "Standard Test Method for Bulk Density ("Unit Weight") and Voids in Aggregate". Moisture of the aggregate at the time of mixing is important, absorption must be satisfied by conditioning the stockpile as necessary to achieve a saturated surface dry (SSD) condition. Otherwise, a dry aggregate can result in a mix that lacks workability for placement and compaction. Conversely, excessively wet aggregates may contribute to paste drainage, causing intermittent clogging of the intended void structure.

Cement: the main binder is Portland cement according to ASTM C150/C150M "Standard Specification for Portland Cement", C595/C595M "Standard Specification for Blended Hydraulic Cements" or C1157/C1157M Standard Performance Specification for Hydraulic Cement.

Where fly ash, blast furnace slag or silica fume are used, they must meet the requirements of

ASTM C618 "Standard Specification for Coal Fly Ash and Raw or Calcined Natural Pozzolan for Use in Concrete", C989 "Standard Specification for Slag Cement for Use in Concrete and Mortars" and C1240 "Standard Specification for Silica Fume Used in Cementitious Mixtures", respectively.

Water: The same requirements apply as for conventional concrete. Recycled water may be used only if it complies with ASTM C94/C94M "Standard Specification for Ready-Mixed Concrete" or AASHTO M-157 "Standard Specification for Ready-Mixed Concrete".

Admixtures: Admixtures shall meet the requirements of ASTM C494/C494M "Standard Specification for Chemical Admixtures for Concrete".

The gradations described can be equated to gradations given in IRAM standards, ASTM D448 "Standard Classification for Sizes of Aggregate for Road and Bridge Construction" and ASTM C33 "Standard Specification for Concrete Aggregates" are analogous to IRAM 1627 "Agregados. Granulometry of aggregates for concrete" and IRAM 1531 "Coarse aggregate for cement concrete. Requirements and test methods" and IRAM 1512 "Fine aggregate for cement concrete. Requirements", respectively.

Other equivalences to national standards are listed below, as there are ASTM analogous standards in Argentina.

- For the determination of the PUV is IRAM 1548 "Aggregates. Determination of bulk density and voids" analogous to ASTM C29.
- The national standard that establishes the definition and requirements for cements for general uses is IRAM 50000 "Cements. Cements for general use. Composition and requirements".
- The one that establishes the requirements for water is IRAM 1601 "Water for cement mortars and concretes".
- The national standard that establishes the requirements for chemical admixtures for concrete is IRAM 1663 "Cement concrete. Chemical admixtures".

2.4. Method of dosage.

As in the case of other concretes, both conventional and with special properties, there are several methods for the design of draining concretes, in this work only the method corresponding to the ACI-522R-10 "Specification for Pervious Concrete Pavement" established by the American Concrete Institute (ACI) [23] will be developed.

For this method, the term "draining concrete" describes mixtures with virtually zero slump and open structure consisting of portland cement, coarse aggregates and little or no presence of fine aggregate, chemical admixtures and water. The combination of these ingredients will produce a hardened material with connected pores, ranging in size from 2 to 8 mm, which allows water to pass through easily. As is characteristic of the designs proposed by the ACI, the method definitions are based on a substantial series of experiments. In this way, relationships are established between the proportions of the mix, the characteristics of the materials and the resulting properties of the mix.

In order to perform the dosage using this method, some of the following required parameters must be established:

A) Compaction: Good or Light.

B) Void content.

C) Water/cement ratio.

D) Coarse aggregate: Size, Unit Weight, Specific Gravity and Absorption data must be available.

E) Fine aggregate.

The first step is to choose the percentage of fine aggregate to be used and a size of coarse aggregate, with this data the ratio b/b0 is selected from the table "6.1. Effective values of b/b0" provided by the standard, where b is the solid volume of coarse aggregate in a unit volume of concrete and b0 is the solid volume of coarse aggregate in a unit volume of coarse aggregate. Then this parameter is used to affect the unit weight of the coarse aggregate, so:

$$P_{agr} = Peso\ Unitario\left(\frac{kg}{m^3}\right).\left(\frac{b}{b_0}\right).1[m^3]PESO\ SECO\ DE\ AGREGADO\ GRUESO \quad (Ecuación\ 1)$$

In the next step, the previously obtained Pagr is adjusted to saturated weight dry surface using the absorption percentage.

$$P_{agr-ssd} = P_{agr}[kg].(1 + \%Abs) \rightarrow PESO\ SSD\ DE\ AGREGADO\ GRUESO \quad (Ecuación\ 2)$$

The volume of paste, in percentage, is then determined using Figure 6.3 provided by the standard, which is entered with the required void ratio until the chosen compaction curve (Good or Light) is reached.

The volume of paste for 1 [m³] of concrete is then obtained:

$$V_{pasta} = 1[m^3].\left(\frac{\%Pasta}{100\%}\right) \rightarrow VOLUMEN\ DE\ PASTA \quad (Ecuación\ 3)$$

With the volume of paste obtained by Equation 4, the cement content "C" is determined using Equation 6 - 1 provided by the standard.

$$V_{pasta} = \frac{C}{\rho_{cemento}} + \frac{\frac{A}{C}.C}{\rho_{agua}} \quad (Ecuación\ 4)$$

Where: ρ = density, in $\left[\frac{kg}{m^3}\right]$

A = water content, in [kg].

C = cement content, [kg].

A/C = chosen water/cement ratio.

Once "C" is obtained, the water content is determined, as well as the volume of aggregates, cement and water and finally the volume of solids:

$$A = \frac{A}{C}.C \rightarrow CONTENIDO\ DE\ AGUA \quad (Ecuación\ 5)$$

$$V_{agr} = \frac{P_{agr-ssd}}{\rho_{agr}.C} \rightarrow VOLUMEN\ AGREGADO\ GRUESO \quad (Ecuación\ 6)$$

$$V_{cemento} = \frac{C}{\rho_{cemento}} \rightarrow VOLUMEN\ CEMENTO \quad (Ecuación\ 7)$$

$$V_{agua} = \frac{A}{\rho_{agua}} \rightarrow VOLUMEN\ AGUA \quad (Ecuación\ 8)$$

$$V_{solidos} = V_{agr} + V_{cemento} + V_{agua} \rightarrow VOLUMEN\ TOTAL\ DE\ SOLIDOS \quad (Ecuación\ 9)$$

The next step is to determine the void volume to verify the initial value adopted, considering a total volume Vtotal of 1 [m^3].

$$\%V = \frac{(V_{total} - V_{solidos})}{V_{total}}.100 \rightarrow VOLUMEN\ DE\ VACIOS \qquad (Ecuación\ 10)$$

Finally, the estimated porosity is verified using *Figure 6.1* provided by the standard, from which the infiltration rate is estimated.

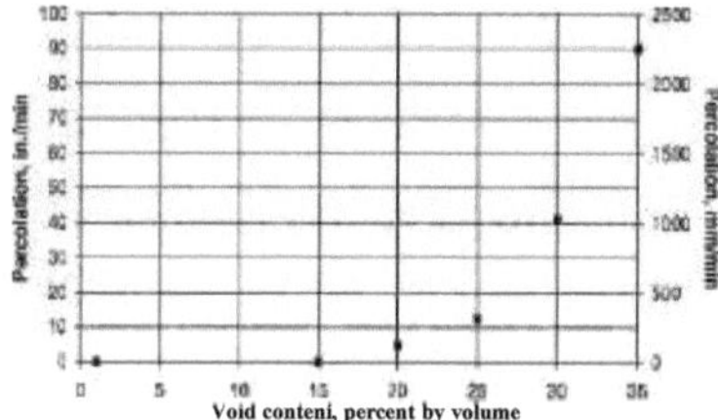

Figure N° 31: Minimum void content for percolation in NAA-NRMCA tests. Figure 6.1 ACI standard 522R-10 [23].

2.5. Uses.

Drainage concrete is currently used for a wide variety of applications, including the following:

<u>Permeable paving:</u>

When it rains the car causes an unpleasant spray mist at the rear, which can affect the visibility of the driver behind and when the water does not flow fast enough on the road the dangerous effect of *aquaplaning* can occur. To avoid this, water-permeable road surfaces made of draining concrete have been developed.

In this case, no water film is formed on the road surface and rainwater can infiltrate on site with a suitable concrete drainage base layer, thus decreasing the risk induced by a wet roadway. In addition, roads and paths can be constructed without superelevation (hydraulic superelevation), leaving the draining concrete surface flat and sloping the lower impermeable layer, which will lead the water away from the pavement structure.

Figures N° 32 and 33: Construction of a permeable concrete road [25].

Figure N° 34: Demonstration of the infiltration capacity of draining concrete [22].

Water purification:

Land development has a negative impact on the quantity and quality of surface runoff that is delivered to natural streams. When natural surfaces (such as lawns and trees) are replaced by impervious structures such as car parks and streets, the water-holding role of soil and vegetation is lost, leading to flooding, erosion damage to runoff channels, decreased groundwater recharge and ecosystem degradation. These same impermeable structures can transport the many pollutants deposited in urban areas, such as nutrients, sediments, bacteria, pesticides, chlorides, hydrocarbons, etc. [26]. [26].

These pollutants create an environmental problem in urban lakes, rivers and wetlands near large cities. Water purification using draining concrete occurs by contact oxidation between gravels, where the biota formed on the inner surfaces of the continuous voids provide an additional biopurification function [27].

It also acts as a filter, allowing water to pass through its structure, but preventing solids larger than its pore size from entering drainage structures or natural soil, acting as a gravity filter.

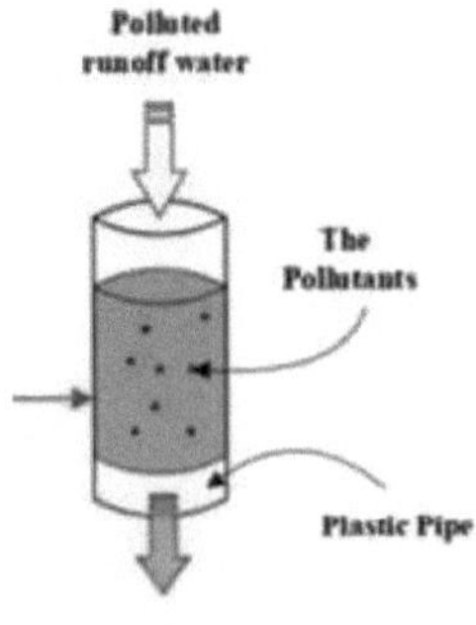

Figure N° 35: Schematic view of a permeable concrete specimen and the run-off water treatment procedure [28].

Noise absorption:

Draining concrete pavements also significantly reduce tyre rolling noise and can contribute to noise protection in cities, on roads and motorways. A decrease of up to 5 [dB] has been detected.

Figure N° 36: Sound equipment located for sound absorption testing [22].

Benches:

To obtain improved shoulders, using less material and therefore at lower cost, drainage concrete can be used. This will not accumulate water, providing a surface with sufficient friction at all times.

Figure N° 37: Road with draining concrete shoulder. Pilot project in Aatal, Münster [22].

Drainage of large surfaces:

In areas such as car parks and impermeable plazas, a large amount of water can accumulate during rainfall, forcing the structure to be designed with slopes to channel the water, then drains, etc. It also decreases surface friction, causing inconvenience to vehicles and pedestrians.

The following picture shows how the surface built with draining concrete (below) remains dry and free of water accumulation.

Figure N° 38: Parking with pervious concrete and plain concrete [29].

Reduction of surface runoff and aquifer recharge:

The decrease in the flow rate is due to the phenomenon of retention, which is the ability to store moisture on the surface to be removed by evaporation.

Aquifer recharge is determined by infiltration capacity, which determines whether the pavement material allows water to pass through its structure and percolate into the ground. If this is done with the objective of reducing the water released into the sewer system, it also reduces surface runoff, slowing peak flow and reducing the likelihood of flooding.

Infiltration and retention are opposing properties, so it is necessary to design the mix with a good relationship between the two [18].

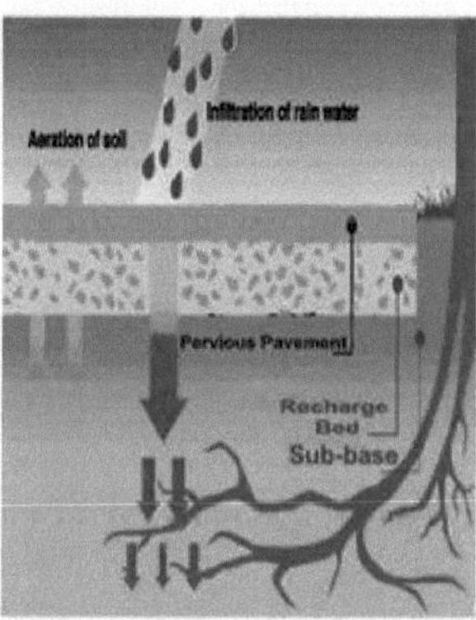

Figure N° 39: Water infiltration mechanism for groundwater recharge [18].

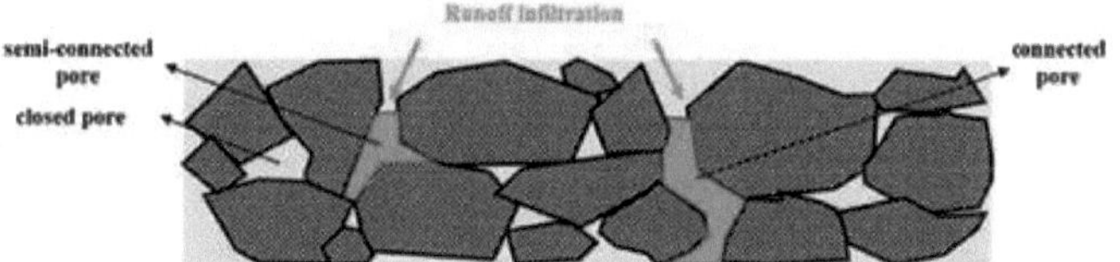

Figure N° 40: Internal structure of draining concrete, it can be seen that water can be retained in closed pores [18]. (The liquid is also retained by surface tension when it adheres to the surface of the material, which, having a large specific surface area, stores a large amount of water).

3. *Key characteristics for the intended use.*

The structure proposed for this project aims to limit the amount of solids entering the sewers while allowing precipitated water to pass through during the design storm. The determining properties for this purpose are the characteristic and maximum pore size, by means of which it is possible to know what size of solids can be retained, and the permeability or infiltration rate, a factor that indicates the infiltration capacity of the structure. The infiltration rate should be indicated, as higher porosity does not ensure higher permeability, since permeability is a function of the area of voids, the size of the voids and their distribution [30].

With these three parameters, it is possible to design the structure and know how it will perform when it is evaluated.

Once the characteristics of the pervious concrete to be used and the intended use of the material have been defined, the terms "drainage concrete" and "pervious concrete" will be used interchangeably from now on.

4. *Mixture to be used developed by CECOVI.*

The values are obtained from laboratory tests carried out by the Centro de Investigación y

Desarrollo para la Construcción y la Vivienda (CECOVI) of the Universidad Tecnológica Nacional, Facultad Regional Santa Fe [12]. In addition to the following data, mixtures with void contents varying between 17.6 and 41% and with compressive strengths between 2.4 and 17.8 [Mpa] at 7 [days] and between 3.5 and 19.8 [Mpa] at 28 [days] have been achieved.

Final Dosage		
COMPONENTS	VALUE	UNIT
Water/Cement Ratio (A/C)	0,35	-
Cement	362	kg
Water	127	kg
Coarse aggregate (PG 3-9)	1567	kg
Fine Aggregate	0	kg
Theoretical Void Volume (V_{vi})	20	%
PUV Theoretical	2056	kg/m3

Table N° 4: Dosage of the mixture used.

Test tube	a/c	Compressive strength	
		7 days	28 days
		Mpa	Mpa
FA	0,35	7,15	9,66
FB	0,35	6,5	8,25

Table N° 5: Simple compressive strength values.

Test tube	a/c	Width (b)	High (h)	Divisions (d)	P=C.d	P=C.d	M	Modulus of rupture
		m	m		kg	MN	Mpa	Mpa
FA	0,35	0,1515	0,1535	870	1,331,100	0,013	1,69	1,89
FB	0,35	0,1525	0,1535	1090	1,667,700	0,017	2,1	

Table N° 6: Results of bending tensile strength.

Test tube	Apparent Volume	Dry bulk density	Volume of voids	Percentage empty	Absorption	Average percentage empty	Average absorption
	cm3	g/cm3	cm3	%	%	%	%
FA	3523	1,857	875,75	19,91	6,49	18,33	6,44
FB	2786,5	1,917	560,36	16,74	6,39		

Table N° 7: Determination of void content and absorption.

Test tube	Pi	Pf	t1	t2	t3	t4	Average	K
	g	g	s	s	s	s	s	cm/s
FDM1	1470,5	1786	6,69	7,09	6,56	6,88	6,92	1,49
FDP1	2015,5	2064,65	7,75	7,28	8,12	7,94	8,67	1,31

Table N° 8: Permeability determination.

From the above tables, the data of compressive strength, coarse aggregate size, permeability, density, modulus of rupture and water-cement ratio will be extracted as reference, which will be taken into account when designing the reception structure concerning the project (together with other values), and will also be used to perform the necessary calculations to know the performance of the structure and to define the numerical models analysed.

Chapter V

Module design

1. Current structure.

Given that the current catchment structure on the Santa Fe pedestrian walkway is 20 cm wide and 30 [cm] deep in its lowest section, increasing according to the direction of the slope, we will seek to design a structure that adapts to this geometry, in order to carry out the least possible intervention, reducing costs and construction time. The aim is to design a structure that adapts to this geometry, in order to carry out the least possible intervention, reducing costs and construction time.

Figure N° 41: View of drainage gratings, Santa Fe pedestrian walkway.

2. Intervention alternatives with draining concrete.

To replace the existing gratings, a thin slab of draining concrete covering the entire surface of the channel could be placed, but given the low tensile strength and the impossibility of placing reinforcement, this option is ruled out. Reinforcements cannot be placed as they would be exposed to air and water due to the presence of interconnected voids in the structure of the material, which would cause them to deteriorate. While the option of using protected reinforcement, stainless steel or other material, could be considered, the bond would not be sufficient to transfer the stresses from the concrete to the steel and vice versa. This is due to the low amount of cement paste, sufficient only to cover the aggregates, which would leave the steel bonded only at the points of contact with the aggregates.

Another option would be to place drainage concrete over the entire volume of the existing channel. This would imply a horizontal runoff of water within the matrix of the material, so that a permeability in the horizontal direction sufficient to deliver the necessary flow would have to be guaranteed. Another disadvantage of this alternative is clogging, as by completely filling the space with draining concrete, the lower face would be in contact with the soil of the channel (impermeable) allowing the continuous accumulation of sediments, gradually decreasing the permeability. Finally, it is noted that the cost would be high, given the volume of material to be used and the quality control that would have to be carried out to ensure that the entire structure meets the requested requirements.

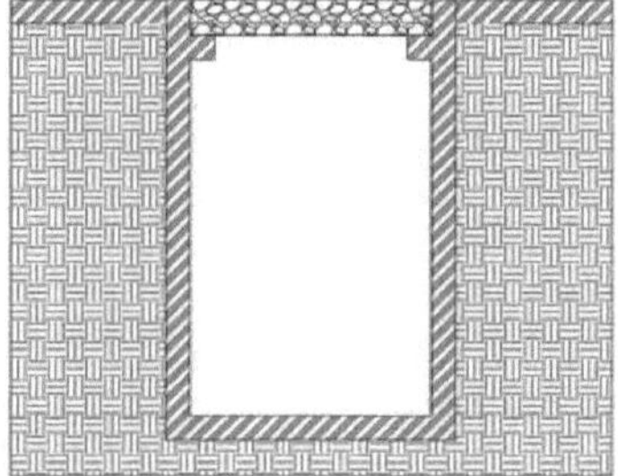

Figure N° 42: Alternative thin slab (discarded). Figure N° 43: Alternative full volume with draining concrete (discarded).

The optimum solution is an isosceles trapezoid section, with a width of 28 [cm] on the upper face, 20 [cm] on the lower face and a height of 20 [cm]. This shape is adopted so that it can be easily placed without the need for support structures, being supported by the walls on its sides, reducing its bending stresses and ensuring that the main stresses are compressive. (The mechanical analysis of the module is developed in chapter VII).

Since the channel has a minimum depth of 30 [cm], a height is obtained where the water can flow freely with the slope given by the channel surface, avoiding the need to guarantee permeability in the horizontal direction. This lower space also allows solids smaller than the pore size to be removed, without being trapped in the internal structure of the material. On the other hand, the volume of material used is reduced, decreasing the costs associated with construction.

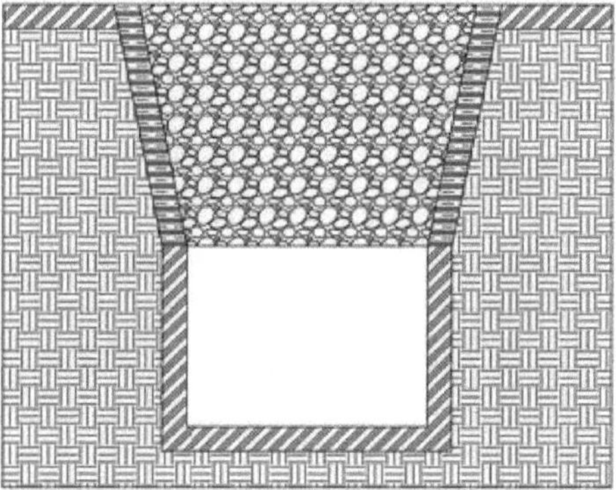

Figure N° 44: Alternative isosceles trapezium section (chosen).

3. *Modulation.*

When a structure is prefabricated, the operations on site are essentially assembly rather than on-site construction and, in most cases, this means an increase in quality, a reduction in waste, improvement and safety on site.

In order to reduce construction time, facilitate construction, avoid on-site formwork and guarantee the properties of the material, prefabricated drainage concrete modules are designed. These modules are placed successively to form the desired structure. Modulation has the added advantage of facilitating the cleaning, movement and replacement of parts of the structure as required at the use and/or construction site stage.

One of the main characteristics into which prefabricated components and systems are divided is the weight of the elements, which are classified as follows:

- Lightweight: The elements weigh less than 100 kg, and are placed manually by one or two operators.
- Semi-heavy: The elements weigh between 101 and 500 kg. They are put into place using simple mechanical means based on pulleys, levers, winches and barges.
- Heavy: The elements weigh more than 500 kg, therefore heavy machinery such as large cranes are required for their installation.

In order to avoid the use of mechanical equipment, modules with a total weight of less than 100 [kg] are designed. Considering the limit values for the safe manual lifting of loads according to Resolution SRT 295/03, of the Ministry of Labour, Employment and Social Security, which indicates that without mechanical aids, the distributed weight should not exceed 25 $\left[\frac{kg}{persona}\right]$ [31] and projecting that two people can move the modules, the module should not weigh more than 50 [kg].

Taking into account that the specific weight of the material to be used is $Pe = 1{,}917 \left[\frac{g}{cm^3}\right]$ (table number 7), the length of the module is set at 0.5 [m], giving a total weight of 0.5 [m]:

$$P = Pe * V = 1{,}917\left[\frac{g}{cm^3}\right] * 24000\ [cm^3] = 46008\ [g] = 46{,}008\ [kg]$$

The weight distributed over two persons gives $P = 23{,}004 \left[\frac{kg}{persona}\right]$, which respects the limit given by the standard. The module is shown in figure 42.

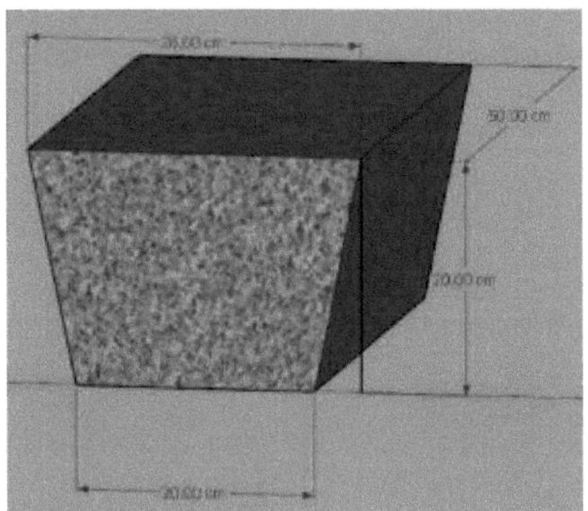

Figure N° 45: Diagram with dimensions of the designed module.

The next chapter deals with the analysis of the discharge and filtration capacity of the proposed structure with the designed modules.

Chapter VI

Analysis of infiltration and filtration capacity of solids/pollutants

1. Design storm.

In order to assess the effectiveness of the draining concrete modules in allowing infiltration of precipitated water, an analysis must be carried out to identify the maximum flow rate to be delivered. This can be done by determining a design storm, which is the most intense hypothetical rainfall event, statistically foreseeable, for a given duration and assigned recurrence [32].

The design storm is defined by the parameters of duration, intensity and recurrence. These values, which are characteristic for each region and climate, can be established from historical records obtained from constant measurement at weather stations.

The recurrence adopted by the Secretariat of Water Affairs and Risk Management of the City of Santa Fe for the verification of the storm drainage system is 5 years [32]. From the data provided by this secretariat, all the values will be obtained to define the design storm to be used to determine the maximum flow that the system will have to deliver.

Weather	Intensity
[min] [min] [min] [min] [min] [min] [min] [min] [min	[mm/h] [mm/h
5	186.1
10	145.4
15	121.6
20	105.6
25	94.1
30	85.2
35	78.2
40	72.5
45	67.7
50	63.7
55	60.2
60	57.2

Table N° 9: Precipitation intensity for different storm durations in the city of Santa Fe with a 5-year recurrence [34].

As can be seen in table 9, adopting a storm duration of 5 [min], the highest intensity of 186.1 [mm/h] is obtained. These values are taken because if the module can deliver the flow produced by the highest intensity storm, the flows for lower intensities can also be delivered. Precipitation falls over a certain surface area, generating a peak flow for the highest intensity. The model is suitable for a sub-basin belonging to the central area of the city of Santa Fe, where the impermeability factor is 100%, so that all the precipitation is transformed into surface runoff.

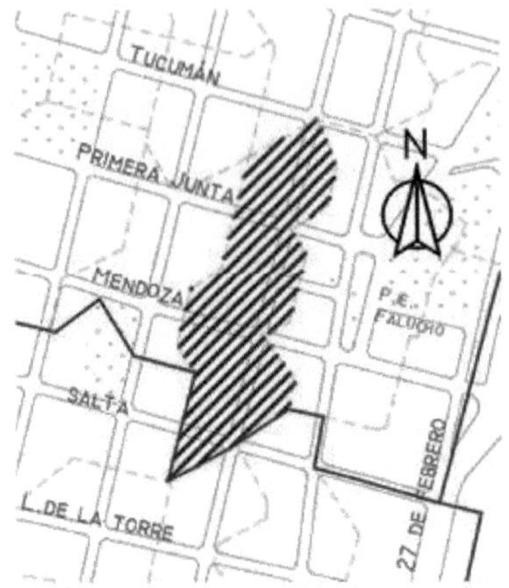

Figure N° 46: Diagram of the sub-basin to be intervened [13].

Considering strips of 1 m in length and given that the total width between building lines is 13 m, the peak flow to be delivered per linear metre is

$$Q_i = i.\ s\ (\text{i})$$

Where:

Q is the flow rate at $\left[\frac{m^3}{s}\right]$.

i the intensity in $\left[\frac{m}{s}\right]$.

S the surface area in $[m]^2$.

$$Q_i = 5{,}1694.10^{-5}\left[\frac{m}{s}\right].13[m].1[m] = 6{,}72.10-4\left[\frac{m^3}{s}\right]$$

2. *Permeability of the module.*

According to data provided by the permeable concrete research group of CECOVI of the National Technological University of Santa Fe Regional Faculty, for a permeable concrete with a water-cement ratio of 0.35, coarse aggregate (PG 3-9), 0% fine aggregate and a theoretical void volume of 20%, the permeability is 1.31 $\left[\frac{cm}{s}\right]$ (table 8).

Since the discharge capacity of the module is given by the permeability and the discharge surface of the module. Where *p* is the permeability 0.0131 $\left[\frac{m}{s}\right]$ and the surface area is the width of the lower section multiplied by the length (a unit length is considered for the calculations so that the values can be compared with those obtained in the calculation of the flow rate, value obtained per linear metre).

$$Qs = p.\ S\ (2)$$

$$Q_s = 0{,}0131\left[\frac{m}{s}\right].0{,}2\ [m].1[m] = 2{,}62.10^{-3}\left[\frac{m^3}{s}\right]$$

The flow rate obtained in equation number **(2)** corresponds to the volume of water that the draining concrete to be used can pass through the structure per unit of time and length.

Comparing the values of equations **(1)** and **(2)**, it can be seen that the discharge capacity of the designed module, with the chosen draining concrete mix and a free evacuation surface of 20 [cm], is greater than the flow produced by the maximum intensity of the design storm. Therefore, it is feasible to use the modules in the stormwater inlets of the Santa Fe pedestrian walkway.

A three-dimensional model of the module is then made in finite volumes and a simulation of the design rain will be made, observing the behaviour of the module with greater detail and safety, being able to evaluate pressures, flow velocities and streamlines in the internal structure, being able to analyse the behaviour of the fluid.

3. *Numerical modelling.*

The problem to be modelled corresponds to a flow in a permeable medium, many softwares have developed a solution to this type of problem and some have used the open platform OpenFOAM, which will be used in the present work.

Modelling and understanding flow in an unsaturated or saturated medium is an important problem in a wide range of scientific domains, such as environmental engineering and groundwater hydrology. A two-phase flow in a porous medium can be modelled by solving the mass conservation equation for each phase, where the phase velocities are expressed using Darcy's law in general. However, a classical and commonly used approach is to neglect the pressure gradient in the unsaturated (typically air) phase to reduce the two-phase flow to one equation, typically the so-called Richards equation [35].

$$\left(\frac{\partial \theta}{\partial t}\right) = \frac{\partial}{\partial x}\left[K(h)\left(\frac{\partial h}{\partial x} + cos\alpha\right)\right] - S \quad \textbf{(3)}$$

Where h is the potential matrix of water in the soil, θ is the volumetric water moisture, t is the time, x is the spatial coordinate, S is the degree of saturation and α is the angle between the flow direction and the vertical axis. The other parameters, equations and calculations are described in the following section.

4. *Analysis using the finite volume method.*

The finite element method (FEM) is a method for representing and evaluating partial differential equations in the form of algebraic equations. In this method, the volume integrals in a partial differential equation containing a divergence term are converted into surface integrals, using the divergence theorem. These terms are then evaluated as fluxes on the surfaces of each finite volume. Since the flux entering a given volume is identical to the flux leaving the adjacent volume, these methods are conservative.

The procedure for creating the model in finite volumes begins with the generation of the geometry and mesh in Salome (Salome 2021), which allows the domain to be discretised and the faces to be defined on which the boundary conditions are assigned.

The first step is to define the geometry of the body, which corresponds to the designed module (measurements in figure 42). Once the geometry has been created, the mesh is created and its discretisation criteria are established.

A discretisation in hexahedra is chosen, as the geometry allows working with regular elements with such volume shapes. When discretising problems in two and three dimensions, finite volumes of poor aspect ratio such as elongated elements should be avoided, examples of which are illustrated in figure 47.

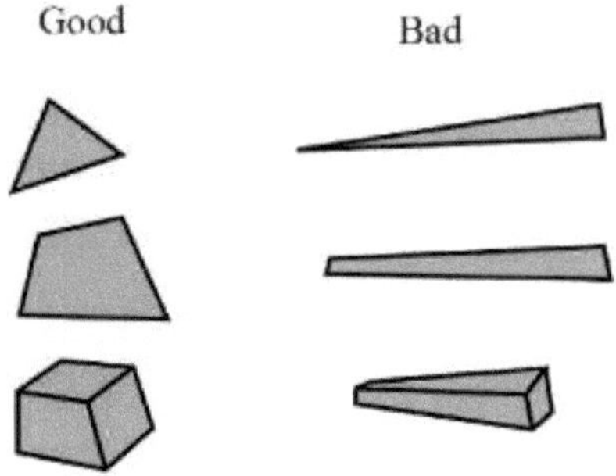

Figure N° 47: Elements with good and bad aspect ratios [36].

As a rough guide, elements with aspect ratios greater than 3 should be viewed with caution and those exceeding a ratio of 10 should be avoided. These elements will not necessarily produce bad results, as this depends on the loads and boundary conditions of the problem, but they do introduce potential problems [36].

Therefore, care is taken to ensure that the elements have a certain uniformity and regularity in order to avoid the results being altered by disturbances due to distorted or irregular elements.

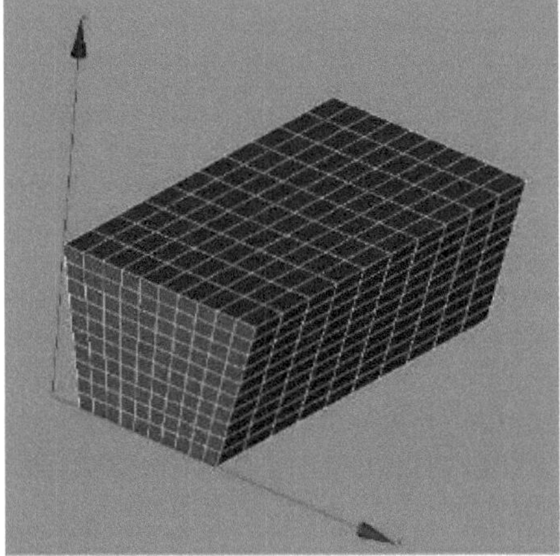

Figure N° 48: Finite volume mesh in Salome 2021.

As can be seen in the figure above, the volumes have a certain uniformity, the distribution is uniform, there are no areas of distortion or elongated elements.

This mesh is then exported for the problem to be calculated in OpenFOAM, using the porousMultiphaseFoam module, a module developed for any two-phase flow (saturated and unsaturated phase) in porous media, is extended to the specific case of the Richards equation which neglects the pressure gradient of the unsaturated phase.

OpenFOAM (Open Field Operation and Manipulation) is a free and open source CFD software, released and developed mainly by OpenFoam Ltd since 2004, which was born at Imperial College London. OpenFoam is organised into a set of C++ modules that make it possible to solve problems ranging from complex fluid flows involving chemical reactions,

turbulence and heat transfer, to acoustics, solid mechanics and electromagnetics [37].
In order to solve the given problem, certain conditions must be specified at the extremes or edges of the independent variable of the equation, as well as an initial value of the problem for all specified conditions. An edge condition or edge value is a datum that corresponds to a minimum or maximum input, internal or output value for a system or component [38].
From the observation of the real behaviour of the system, the boundary conditions that govern the problem are observed and the groups of faces are created in order to enter them into the model, so that the model behaves in a way that is approximately the same as the real problem.
In this case, the values of pressures (h), flow velocities (Utheta) and flow conditions on each face of the model must be determined, as detailed below.
The pressure head on the upper face is equal to zero (h1 = 0 [m]), since as demonstrated above, the module has the capacity to deliver the peak flow for the design storm, so there is no accumulation of water to generate a pressure head. On the other hand, on the lower face the pressure will be equal to zero (h2 = 0 [m]), since the free flow of water towards the drain is allowed, so there are no static pressures.
The flow velocity is calculated by dividing the total flow produced by the design storm at the contributing surface by the discharge area of the module:

From equation **(1)** it is obtained that the flow rate is: $Q = 6{,}72.10 - 4 \left[\frac{m^3}{s}\right]$
The free discharge area of the module being equal to: 5 = 0,2 [m] * 1 [m] = 0,2 $[m]^2$
The speed will be equal to:

$V = Utheta = \frac{Q}{S} = \frac{6{,}72.10-4 \left[\frac{m^3}{s}\right]}{0{,}2\ [m^2]} = 3{,}36.\ 10^{-4} \left[\frac{m}{s}\right]$, orresponding to an intensity of rainfall of 186.1 $\left[\frac{mm}{h}\right]$ and a roadway of 13 [m].
This value is imposed as a boundary condition on the top face, with the velocity in the vertical direction, with a negative (up-down) direction, representing the water inflow from the design storm.
A zero flow is established on the lateral support faces, as they will be supported on material considered impermeable, so that the water flow will be directed towards the lower face.
The "empty" condition applies on both cross-sides, as there is no horizontal flow in the designed model.

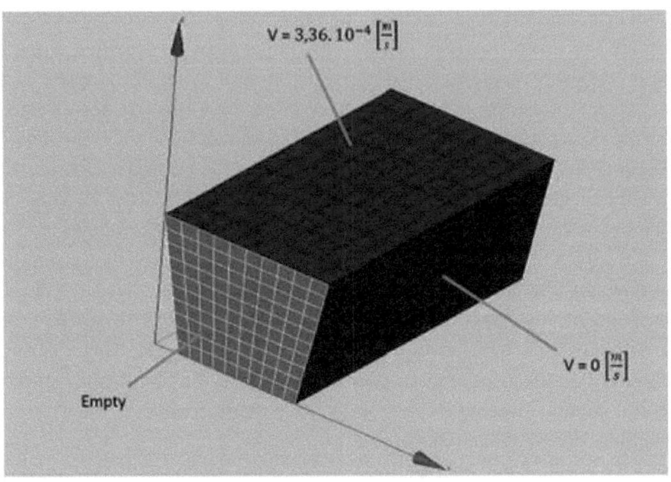

Figure N° 49: Diagram of boundary conditions.

Finally, the parameters concerning the material, the behaviour of the system and the fluid are detailed. An incompressible, viscous and isothermal fluid is assumed.

The hydraulic conductivity of the material measured in the tests is $1{,}31.10^{-2}\left[\frac{m}{s}\right]$. In this case the software requires the intrinsic permeability k, which is obtained as follows:

$$K = \frac{\rho.k.g}{\mu}\ (2) \rightarrow k = \frac{K.\mu}{g.\rho} \qquad k = \frac{1{,}31.10^{-2}\left[\frac{m}{s}\right].1.10^{-3}\left[\frac{kg}{m*s}\right]}{9{,}81\left[\frac{m}{s^2}\right].1000\left[\frac{kg}{m^3}\right]} = 1{,}34.10^{-9}\ [m^2]$$

Where k is the permeability $(1{,}31.10^{-2}\left[\frac{m}{s}\right])$, P the density of the fluid $(1000\left[\frac{kg}{m^3}\right])$, μ the dynamic viscosity of the fluid $(1.10^{-3}\left[\frac{kg}{m*s}\right])$, k the intrinsic permeability ($[m^2]$) and g the gravity $(9{,}81\left[\frac{m}{s^2}\right])$.

Finally, the Van Genuchten parameters are adopted, by means of which the modulus is analysed in its initial unsaturated condition, looking for the equilibrium point. The saturated matrix water content θs = 0.25, the matrix residual water content θr = 0.025, the inverse capillary length a = 0.145 and the pore size distribution parameter n = 2.68 [39].

The results obtained in paraView, which is an open source cross-platform data visualisation and analysis application, are presented below.

Figure N° 50: Streamlines and velocity magnitude (values in [cm/s] [40].

Figure 50 shows the streamlines for the case analysed, together with the colours corresponding to the velocity scale. The verification of the boundary conditions can be seen in velocities on the contours, as well as the path of the fluid, which moves mainly in a downward vertical direction, increasing its velocity as it crosses the module, not only due to the influence of gravity, but also due to the difference between the inlet and outlet area, the latter being smaller, keeping the flow constant (by conservation of masses), with a smaller area, the velocity increases. The observed velocity values do not represent a risk of erosion or damage to the module [41].

18%	
Velocidad de filtración [mm/s]	Re
0.031	0.765
0.072	1.79
0.133	3.30
0.202	5.02
0.273	**6.78**
0.397	9.86
0.488	12.1
0.623	15.5
0.735	18.3

Figure N° 51: Reynolds numbers for infiltration flow in draining concrete [42].

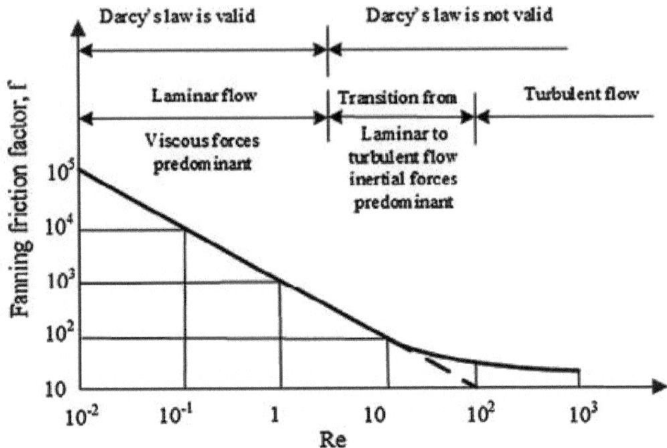

Figure N° 52: Water flow conditions in porous media [42].

Based on the information obtained from Figure 51, which indicates the values of the number of Reynolds for different infiltration velocities in a permeable concrete with 18% effective porosity and from the graph in figure 52, it is observed that, since the maximum calculated velocity is 0.17 [~^~], the flow will be laminar, this value also indicates that Darcy's law is valid to analyse the problem [42].

Figure N° 53: Pressure diagram (values in metres of water column) [40].

Figure N° 54: Pressure diagram together with streamlines [40].

Figure 53 shows the pressure generated by the water flow on the zero flow walls of the module. This pressure is generated by the change in the cross-section of the module, as shown in figure 54, the water is channelled towards the lower area (with a smaller cross-section) through the side walls, producing a pressure of 0.16 [~~⅛], a negligible pressure compared to the strength of the draining concrete used.

Based on the above, the success of the design in meeting the objectives set for fluid delivery can be corroborated, as the module has sufficient permeability to deliver the design storm with the available area.

The computational model shows that the boundary conditions are correctly placed, there are no disturbances in the flow, which is always laminar, a correct distribution of pressures correlated with the direction of flow and the shape of the module, therefore it is understood that the simulation is valid for the case analysed.

The last point in this chapter is the analysis of the module's capacity to filter solids and reduce pollutants.

5. Filtration capacity.

5.1. Filter analysis of the module.

Since the main objective of the project is environmental in nature, reducing the amount of urban waste present in the water discharged from storm drains, this point is essential to evaluate the efficiency of the material and the module with respect to this purpose.

Filtration is the unitary process of separating solids in a suspension through a porous mechanical medium, also called a sieve, screen or filter. In a suspension of a liquid through a porous medium, it retains solids larger than the porosity size and allows liquid and particles smaller than the porosity size to pass through [43].

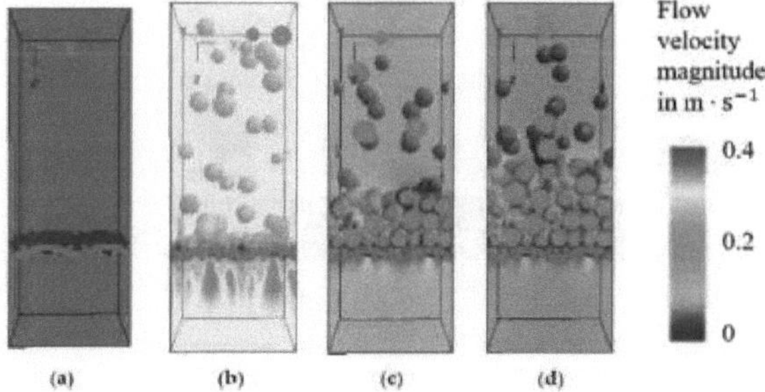

Figure N° 55: CFD-DEM simulation of the filtration process: (a) filter, (b) flow with suspended particles, (c) solids separation, (d) accumulation of objects and cessation of flow [44].

When analysing the filtration capacity, the size of the particles to be removed in the process and their comparison with the pore dimensions of the material used for the filtration process, among other aspects, must be taken into account.

In this case, the aim is to eliminate urban waste from the water, such as bottles, wrappers, bags, packaging, plates, glasses, cutlery and plastic lids; also metals such as cans or lids; glass containers; cigarette butts, etc. The waste that is most frequently deposited on the public highway is considered, discarding the composition of household or industrial waste, only the waste described above is worked on.

The size of the pores in the draining concrete matrix is very difficult to measure, as most of them are located inside the body. Currently the best option to obtain the pore size is to perform a computed tomography of the object in question, with this technology it is possible to obtain information on the internal structure and composition of the body to be analysed, this information is processed by computers and transformed into images of the interior of what has been analysed.

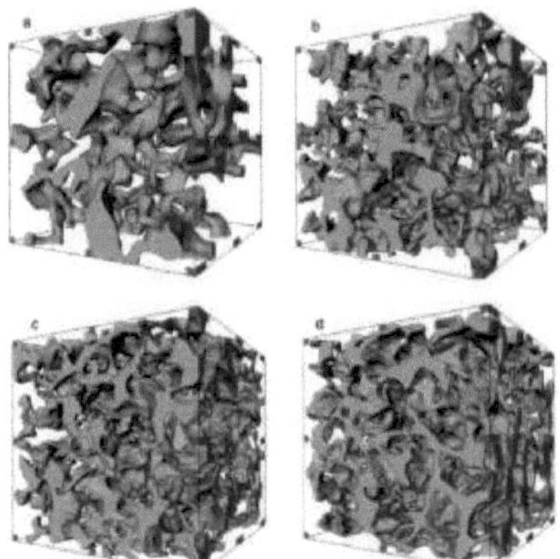

Figure N° 56: Example of a 3D model of interconnected pores of draining concrete [42].

As part of the research project on draining concretes, together with CECOVI, a CT scan was carried out on a test specimen with the dosage described above. This non-destructive test was carried out at the INTI (National Institute of Industrial Technology) located in Rafaela. As a product of the test, a three-dimensional model of the analysed specimen is obtained, which can be observed from any angle, cuts are made in all directions to acquire information on the internal structure and, from a computational analysis, characteristic values of the material can be obtained, such as void volume, distribution factor, average and maximum pore size, interconnection between pores, etc.

The following figure (number 54) shows a view of the CT scan, with the specimen in three dimensions, where the interior voids of the structure, occupied by the coarse aggregate or cement paste, can be clearly distinguished. Examining the 3D tomography also shows the interconnection between the pores, which is a necessary condition for the material to be permeable.

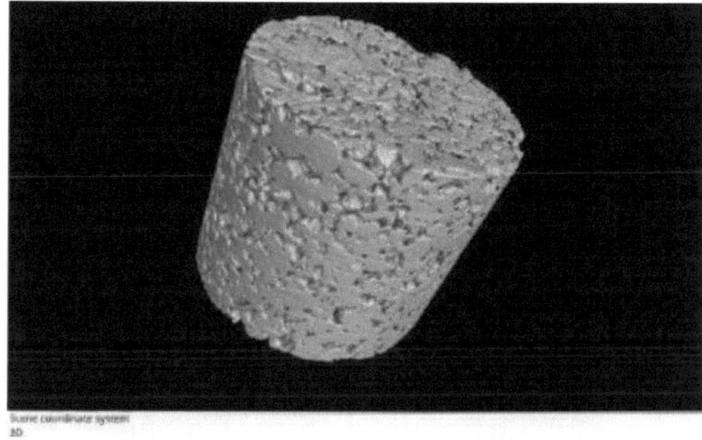

Figure N° 57: 3D structure obtained by computed tomography at INTI Rafaela [40].

Figure 55 shows a cross-section of the specimen, together with a diagram showing where the cut has been made and a scale, which allows the internal distribution of the pores to be observed and the pore size to be analysed.

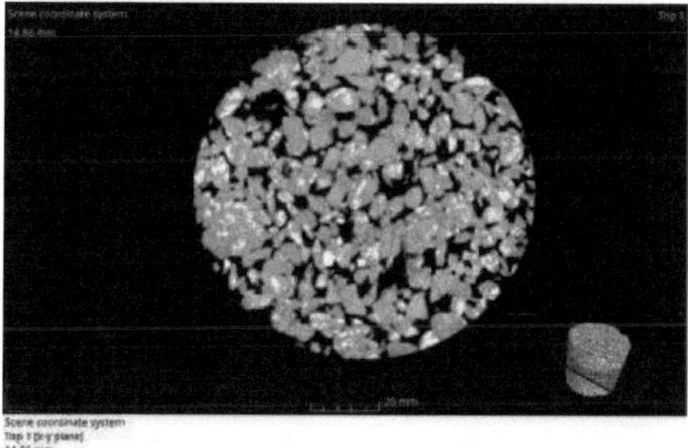

Figure N° 58: Cross-section of the three-dimensional model obtained by computed tomography [40].

Figure 56 represents a longitudinal section of the specimen, while figure 57 represents a rotation about the main axis. With all these view options, the internal structure of the material can be fully examined, allowing all pores to be measured in three dimensions, their interconnection and shape to be observed in detail, whether there are irregularities in the pore distribution or any other relevant aspect.

Figure N° 59: Longitudinal section of the three-dimensional model obtained by computed tomography.

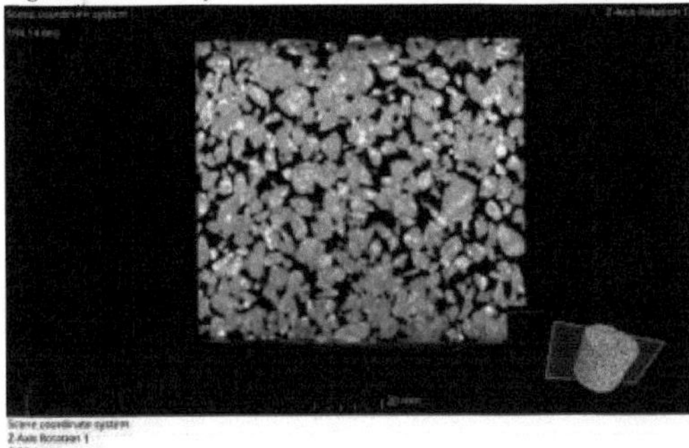

Figure N° 60: Cut in a plane rotating about the principal axis of the three-dimensional model obtained by computed tomography.

After analysing all the information provided by the computed tomography, it was determined that the maximum pore size is 16 [mm] in diameter (largest open dimension), so the module could retain all objects whose dimensions are larger than this value. Since the objective is to remove objects such as bottles, cans, plastic wrappings, etc. from the stormwater and all of them have dimensions larger than 16 [mm], the material to be used fulfils the function of retaining them and separating them from the fluid.

The particles that will pass through the structure will be those that are smaller than the specified dimensions, such as sand, clay, silt or small pieces of any other material.

One of the main environmental problems to be reduced is the presence of microplastics in nature. This problem is present in the Setúbal Lagoon, with a concentration of plastic waste of 100 bottles per square kilometre, and the presence of microplastics was also detected, which would be ingested by fish and birds, accentuating the problem [6]. The following images show the large amount of plastic waste present in the lagoon's riparian zone.

Figure N° 61: Plastic waste on the shore of Setubal Lagoon [6].

Figure N° 62: Plastic waste on the shore of Setubal Lagoon [6].

Microplastics are divided into two types, primary and secondary. Examples of primary microplastics include microbeads found in personal care products or pellets used in industry, primary microplastics enter the environment directly through various channels, such as in the disposal of personal products containing microplastics, unintentional loss during transport, abrasion of tyres on the road or of synthetic fabrics during the washing process, etc.

Secondary microplastics are formed by the breakdown and decomposition of larger plastics, usually when larger plastics are subjected to weathering, e.g. through exposure to wave action, wind abrasion and ultraviolet radiation from sunlight [45].

Secondary microplastics is where this project works by removing large plastic objects (larger than 5[mm] in diameter) from the stormwater and thereby reducing the formation of microplastics in the environment, as they would be removed from the filter for collection and safe disposal by the relevant state agency, along with the waste collected by the current pedestrian sweeping and cleaning service.

5.2. Clogging and clean-up procedure.

From the above it can be noted that after a certain period of time the filter may reduce its efficiency in water infiltration (permeability) by a process of clogging, a process by which the pores of the material become gradually clogged by an accumulation of sediment.

Through the analysis of the results obtained from a study carried out by the Civil Engineering, Materials and Environment Research Group (GIICMA) of the National Technological

University - Concordia Regional Faculty, whose objective was to determine the infiltration rates of clean draining concrete specimens and their variation according to the degree of clogging, it is observed that the complete progress of the clogging phenomenon of the porous pavement due to sediment contributions generates final infiltration rates equal to or lower than 1% of the initial value corresponding to clean samples [46].
Therefore, the following cleaning procedure is proposed for the modules, in order to ensure the permeability of the structure over time.
All modules shall be provided with a slit on one of their cross sides of 10 [mm] on each side, as shown in the following drawing in figure 60, which shows a corner of the module in plan view. The purpose of this is to be able to insert a bar that allows an operator to lift the module on one of its sides and, with the help of another, remove it from its position in order to clean it. This section is adopted as it would not represent a decrease in filtration capacity, as the maximum pore size is 16 [mm] in diameter, exceeding the size of the space designed for the entry of the bar.

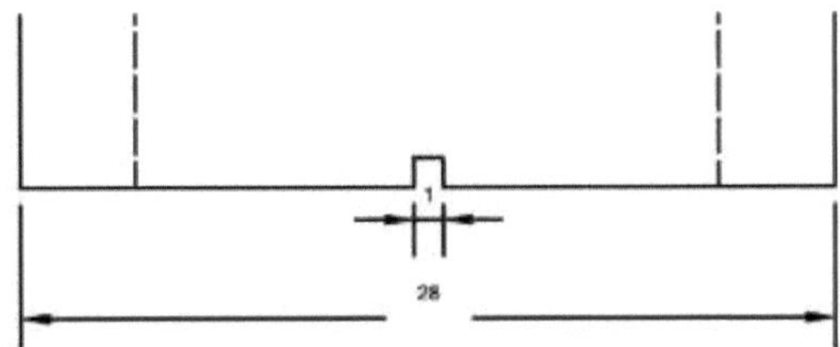

Figure N° 63: Diagram of the module insertion and removal slot (scale in centimetres).

After being removed, the modules are placed upside down with respect to their position in the drainage structure and water is circulated through their internal structure (backwashing), to remove all the residues accumulated inside them and to restore the original permeability of the system. They are then placed in their original position.
The durability of draining concrete is between 6 and 20 years, where the cause for the end of its service life is the clogging of its pores [47]. Therefore, the cleaning procedure should be performed at most once every 5 years to guarantee the design permeability.

5.3. Reduction of pollutants according to studies analysed.

Different studies have demonstrated the great capacity of drainage concrete in the reduction of pollutants in water and air, which are based on different properties of this material, not only filtration, but also biodegradation, chemical reactions, absorption of compounds, etc.
For example, a study conducted by the Key Laboratory of Road and Traffic Engineering, Tongji University in Shanghai, China, indicates that some surface air pollutants, such as solid particles, carbon monoxide, carbon dioxide, hydrocarbons, nitrogen oxides and sulphur oxides, etc. can be removed from the air to improve the atmospheric environment by means of a TiO2 Semiconductor mixed in permeable concrete to prepare photocatalytic pavement. It was found that photocatalytic pavement could react with surface air pollutants to purify the atmosphere [18].
Another research project carried out by the Department of Civil Engineering, Wasit University, Iraq showed that the reduction of COD (chemical oxygen demand) by permeable concrete in water filtration can be approximately 54 %, while the reduction of BOD (biochemical oxygen demand) can reach values of 79 % [47]:
These are examples that show the ability of this material to act as a filter, efficiently removing

pollutants from the water or air by different mechanisms. The aforementioned effects could be added to the positive impacts on rainwater runoff, improving the quality of the water deposited in the natural watercourses as a result of rainfall in the area analysed in the Santa Fe pedestrian area even more than expected due to mechanical filtration.

Chapter VII

Mechanical analysis

1. Determination of the fault model and software to be used.

After having studied the permeability and pore size of the material, determining parameters to evaluate the capacity of the module to deliver the necessary water and filter the solids respectively, a resistance analysis of the system will be carried out.

It should be noted that the location of the project was chosen in a pedestrian zone in order to reduce the mechanical demands on the modules as much as possible (in addition to the high presence of urban waste due to the agglomeration of pedestrians). By locating the structure on the Santa Fe pedestrian walkway, heavy vehicles are prevented from passing over it and the frequency of light vehicles, which only enter in small numbers to enter the few garages located in the area in question, is reduced.

Therefore, the mechanical resistance of the module should be evaluated for the passage of a light vehicle over it, noting that this will be the highest load to which the element will be subjected. The fatigue phenomenon associated with progressive deterioration over time is disregarded, since the structural designs for pavements foreseeing their failure under the concept of fatigue are carried out for the order of 1.10^6 repetitions or more, a value much higher than expected in the useful life of the module, so that the static analysis is decisive compared to the fatigue analysis in this case.

The software chosen to make the computational model is Autodesk's Robot Structural Analysis Professional (2023), which is a structural load analysis software, which allows the creation of designs of plane-deformation structures, design and assembly of buildings or shell structures, The first alternative is chosen because it is better suited to the geometric and load characteristics of the problem in question, since it has an axis of symmetry and the section is constant in one direction, being able to analyse only one plane of the body, disregarding the deformations in the longitudinal direction, which will be irrelevant compared to the deformations in the analysed plane. This simplifies the computational model, reducing complexity and calculation time without generating significant variations in the calculations if longitudinal deformations would be considered. The shape of the module is even coincident with the example scheme proposed by the programme for the design of plane-deformation structures, which indicates the correct adaptation of the geometry in the chosen design alternative (figure 64).

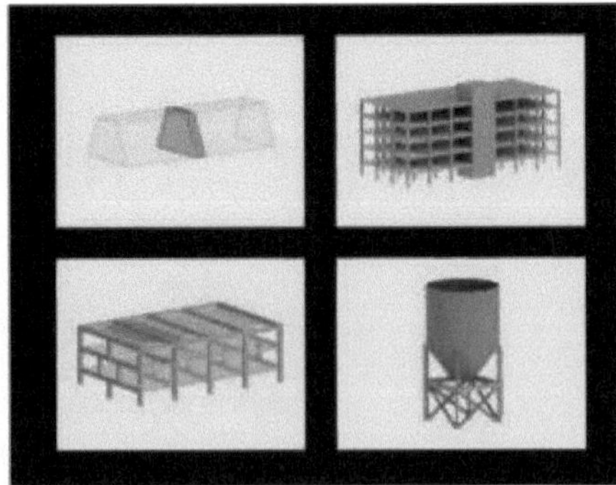

Figure N° 64: Schemes of design alternatives in Robot Structural Analysis Professional (2023).

In order to create a numerical model to simulate the situation of a vehicle tyre resting on a draining concrete module, it is necessary to establish different parameters that characterise the material, the supports or bonding conditions and the load.

2. *Material characteristics.*

The module will be composed entirely of draining concrete, reinforcement cannot be used as it would be exposed to air and water due to the presence of interconnected voids in the structure of the material, which would cause it to deteriorate. While the option of using protected rebars, stainless steel or other material, could be considered, the bond would not be sufficient to transfer the stresses from the concrete to the steel and vice versa. This is due to the low amount of cement paste, sufficient only to cover the aggregates, which would leave the steel bonded only at the points of contact with the aggregates.

The tensile stresses must therefore be resisted in their entirety by the concrete, a material with very low resistance to this type of stress.

The parameters requested by the programme to characterise the material and to be able to carry out the calculations are the following:

Young's modulus: is a mechanical property that measures the tensile or compressive stiffness of a solid material when the force is applied longitudinally. It quantifies the ratio between the tensile/compressive stress σ (force per unit area) and the axial strain ε (proportional strain) in the linear elastic region of a material. The value adopted is 20900 [Mpa], obtained from laboratory tests at the Technical University of Munich [22].

Poisson's ratio: is a measure of the Poisson's effect, the deformation (expansion or contraction) of a material in directions perpendicular to the specific direction of loading. Most materials have Poisson's ratio values ranging from 0.0 to 0.5. The value adopted is 0.467, corresponding to a characteristic value for draining concretes, experimental data, there are not enough laboratory tests to determine an exact value for each dosage.

Shear modulus: is an elastic constant that characterises the change in shape experienced by an elastic material (linear and isotropic) when shear forces are applied to it. Adopted value equals 10720 [Mpa], typical value for H-25 concrete (obtained from software data).

Specific gravity: The specific gravity is the ratio between the weight of a substance and its volume. It is adopted $1857 \left[\frac{kg}{m^3}\right]$ equal to $18,21 \left[\frac{KN}{cm^3}\right]$. This data is obtained from the tests carried out by CECOVI, table number 7 of this work.

Coefficient of thermal expansion: is the quotient that measures the relative change of length or

volume that occurs when a solid body or fluid inside a container changes temperature causing thermal expansion or contraction. Value entered $0,000012 \left[\frac{1}{°C}\right]$, characteristic value for a common concrete (value obtained from the software).

Damping: is an influence within or on an oscillating system that has the effect of reducing or preventing its oscillation. 0.15 is used, which is the characteristic value for a common concrete (value obtained from the software).

Characteristic resistance: is the statistical value of the resistance that corresponds to the probability that ninety per cent (90%) of all test results in the population exceed this value. Value entered, 8.25 [Mpa], (table number 5).

Type of specimen related to the data entered: this refers to the geometric shape of the specimen used in the tests carried out to obtain the characteristic or design strength, the alternatives are cubic or cylindrical. Since the compression tests carried out by CECOVI were performed with cylindrical specimens, this option is included.

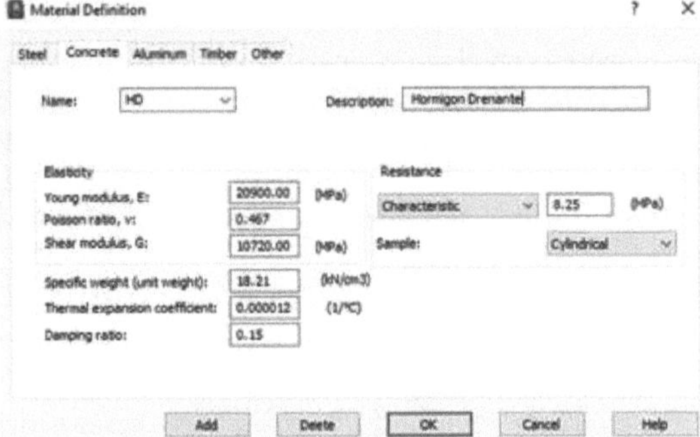

Figure N° 65: Window for material definition in Robot Structural Analysis Professional (2023).

3. *Linking conditions.*

The link conditions are the constraints imposed on each degree of freedom of the structure. As a plane strain analysis will be developed, the degrees of freedom of a cross-section figure of the module will be 3, two displacements and one rotation. For the system to be stable, the linkage conditions must be at least 3, with no apparent linkage.

Since the modules are simply supported on their lateral faces, continuous boundary conditions are assumed on the lateral faces, as shown in the figure below.

Figure N° 66: Representation of the module and boundary conditions Robot Structural Analysis (2023).

A support angle was established for the bonds, corresponding to the direction perpendicular to the face where each bond was applied.

For the simulation of a simply supported bond, a coefficient of friction was established and a value of mi= 0.4 was adopted.

To respect the isostaticity of the system, the fixed link condition in the Z-direction was attached to one of the continuous links mentioned above.

4. *Load analysis.*

4.1. *Contact surface.*

The primary function of tyres is to provide an interface or contact surface between the vehicle and the road surface, transmitting the received load to the ground. The contact patch is that part of a vehicle's tyre that is in actual contact with the road surface.

The tyre contact area for the four wheels of a typical medium-sized car is approximately 8.5 x

11 inches (21.59 [cm] x 27.94 [cm]) [49].

As the tyres carry the vehicle load, it causes the tyres to deflect until the average contact area pressure is equalised with the internal air pressure of the tyres. Assuming a typical passenger tyre inflates to 35 [psi] (0.24 [Mpa]), then a 350 [lb] (158.757 [kg]) load would require an average of 10 square inches (0.0064516 [m^2]) of contact area to support the load. Larger loads require more contact area (more deflection) or higher tyre pressures, or even a larger tyre.

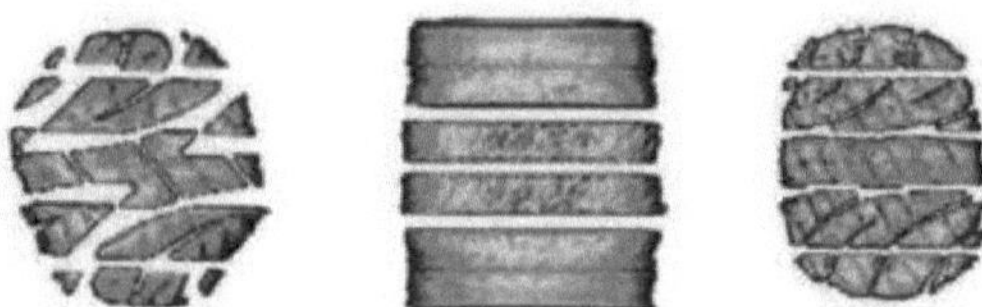

Figure N° 67: Images obtained using smart skin. Sample of three contact traces of different vehicles, with the pressure distribution represented by colour scale [50].

As a support section, an area of 20 [cm] by 20 [cm] sides was established, corresponding to the average support area of a light vehicle [49].

4.2. Tyre load.

The value indicated by the AASHTO 1993 standard for the axle load on a light vehicle is 0.6 [Tn] per axle, i.e. 600 [kg] per axle. This load shall be distributed to each of the tyres, with each tyre receiving the equivalent of 300 [kg] [51].

From this value a linear load is obtained which will be used for the computational model, the value of 300 [kg] is divided by the 20 [cm] depth of the footprint, thus obtaining a linearly distributed load equal to 15 $\left[\frac{kg}{cm}\right]$. To be entered into the program, this value must be converted to $\left[\frac{kg}{cm}\right]$ giving a final value of 0.15. $\left[\frac{kg}{cm}\right]$

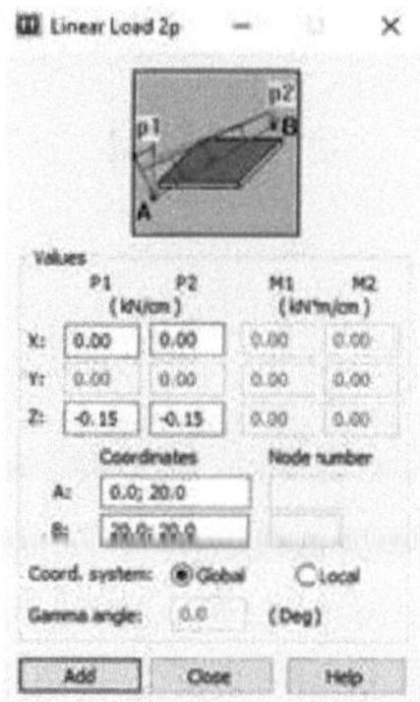

Figure N° 68: Window for entering a linear load.

For the load to be defined correctly, the corresponding value must be entered at each end of the line (points A and B), the negative sign indicates the downward direction in the coordinate system. The points A and B are defined by the indicated coordinates, the position of the most unfavourable load was adopted, i.e. the one that generates the maximum moment on the lower face. This position coincides the mid-section of the footprint with the midpoint of the module.

A schematic of the load location is shown in the following image.

Figure N° 69: Schematic of the computational model with linear load applied.

5. *Results obtained.*

Once the complete model, consisting of geometry, material parameters, bond conditions and loads, is loaded, the program is run to obtain the results and proceed with the analysis. The program displays different colour scale diagrams for stresses and strains, with a choice of detailed x-axis or y-axis stresses, or shear stresses for the xy components, or alternatively the principal stresses can be plotted.

Different diagrams with the results obtained are shown below.

Figure N° 70: Stress diagram analysed in the X direction.

As can be seen in figure 70, negative (compressive) stresses are generated in the upper part of the module, while in the lower face the stress is positive (tensile).

Figure N° 71: Stress diagram analysed in the XY direction (shear stresses).

Figure 71 shows the shear stress diagram, the area with the highest values is close to the supports.

6. Resistance analysis.

In order to know if the module will withstand the stresses it will be subjected to, it is necessary to analyse the principal stresses that develop, which are calculated by the software.

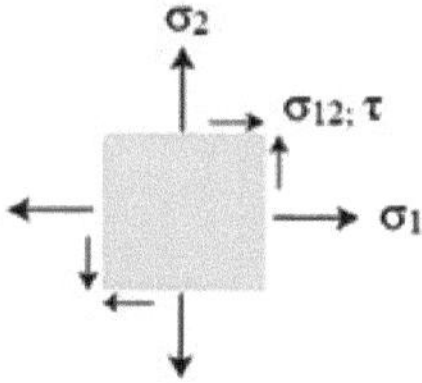

Plane Stress

Figure N° 72: Diagram of principal stresses in a flat element.

The following figure shows the colour scale diagram for the stresses σ_1, the highest positive value is on the lower side and is represented by the colour red, it is equal to 0.02 [Mpa], while the highest negative value appears on the upper side with the colour blue, the latter being equal to -0.02 [Mpa].

Figure N° 73: Principal stress diagram σ_1.

The second main direction σ_2 is also analysed, which gives a maximum value of negative stress on the upper face equal to -0.02 [Mpa].

Figure N° 74: Principal stress diagram σ_2.

Finally, the colour scale diagram for the principal stresses τ is shown, the highest magnitude being at the supports and at the bottom face, taking a value of 0.01 [Mpa].

Figure N° 75: Principal stress diagram τ.

The values obtained should be compared with the strength of the material used, if the stresses are less than the strength, the modulus will be able to withstand the applied loads, otherwise the structure should be rethought.

The equations used to evaluate the resistance correspond to the CIRSOC 201 regulation.

$$Design\ strength\ (\varphi.\ S_n) \geq Required\ strength$$

I.e.: *Resistance. Nominal. Reduction Factor ≥ Load Factor. Serviceability requirements*

The φ = 0.55 is adopted for all cases, based on Article 9.3.5. of CIRSOC 201, which indicates that the strength reduction factor ϕ, for bending, compression, shear and crushing in simple structural concrete, in accordance with Chapter 22, shall be φ = 0.55.

<u>Compression:</u> $\varphi P_n \geq P_u$

$$0{,}55 * 8{,}25\ [Mpa] \geq 0{,}02[Mpa]$$

$$4{,}5375\ [Mpa] \geq 0{,}02[Mpa]$$

$$VERIFICA$$

<u>Traction:</u> $\varphi P_n \geq P_u$

$$0{,}55 * 1{,}89\ [Mpa] \geq 0{,}02[Mpa]$$

$$1{,}0395\ [Mpa] \geq 0{,}02[Mpa]$$

$$VERIFICA$$

<u>Cut-off:</u> $\varphi V_n \geq V_u$

$$0{,}55 * 1{,}89\ [Mpa] \geq 0{,}02[Mpa]$$

$$1{,}0395\ [Mpa] \geq 0{,}02[Mpa]$$

$$VERIFICA$$

It was found that all values verify the stresses to which the module will be subjected, so it can be used with the designed dimensions and material.

Chapter VIII

Material calculations.

1. Calculation of materials required for the modules.

Based on the volume of one of the modules and the dosage used, the quantities of cement, stone, etc. are estimated in order to manufacture all the modules necessary to complete the entire project.

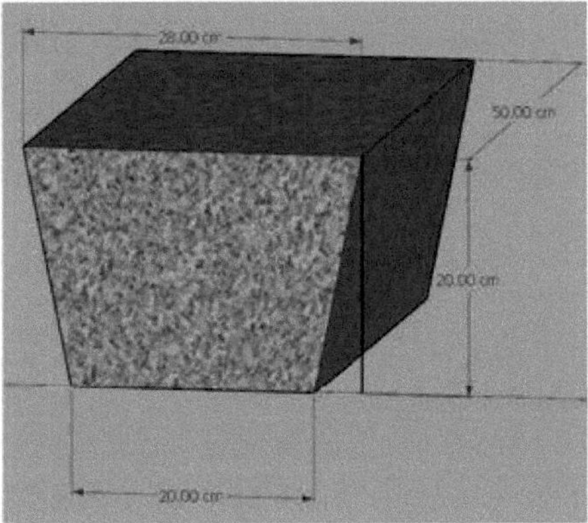

Figure N° 76: 3D schematic of the module with dimensions.

Cross-sectional area:

$$A = \frac{lado\ inferior + lado\ superior}{2} * altura$$

$$A = \frac{20\ [cm] + 28\ [cm]}{2} * 20\ [cm]$$

$$A = 480\ [cm^2]$$

Volume of a module:

$$V_m = A * Longitud$$

$$V_m = 480\ [cm^2] * 50\ [cm]$$

$$V_m = 24000\ [cm^3]$$

Total length of work:

The distances were measured on Google Maps and yielded the following values:

Esquina Salta a calle Mendoza: 124,8 [m] ≈ 125 [m].

Esquina Mendoza a calle Primera Junta: 120,9 [m] ≈ 121 [m]

Corner of Primera Junta to Tucumán street: 124,78 [m] ≈ 125 [m].

Sum: $L = 125$ [m] + 121 [m] + 125 [m] = 371 [m].

Figure N° 77: Diagram of the sub-basin to be intervened [13].

<u>Number of modules required:</u>

$$N = \frac{L}{longitud\ del\ módulo}$$

$$N = \frac{371\ [m]}{0{,}5\ [m]}$$

$$N = 742$$

$$N = 742 * 1{,}05 \approx 780$$

In total, approximately 742 modules would be used. Considering a 5% wastage, it is estimated that 780 modules would be needed.

$$V = V_m * N$$

$$V = 24000\ [cm^3] * 780$$

$$V = 18720000\ [cm^3] = 18{,}72\ [m^3]$$

<u>Total volume of draining concrete:</u>

Approximately 19 [m^3] of drainage concrete will be required. On the basis of the dosage used, the quantity of each of the materials required to produce the drainage concrete is calculated separately.

Final Dosage		
COMPONENTS	VALUE	UNIT
Water/Cement Ratio (A/C)	0,35	-
Cement	362	kg
Water	127	kg
Coarse aggregate (PG 3-9)	1567	kg
Fine Aggregate	0	kg
Theoretical Void Volume (V_{vt})	20	%
PUV Theoretical	2056	kg/m3

Table N° 10: Dosage of the mixture used.

The table detailing the dosage indicates that 362 [kg] of cement and 1567 [kg] of coarse aggregate (PG 3-9) are used for each cubic metre of draining concrete, with a water-cement ratio of 0.35. The following total quantities will therefore be needed to achieve the required modules.

Cement: $362\ \left[\frac{kg}{m^3}\right] * 19\ [m^3] = 6878\ [kg]$

Coarse aggregate (PG 3-9): $1567\ \left[\frac{kg}{m^3}\right] * 19\ [m^3] = 29773\ [kg]$

In summary 6878 [kg] of cement will be required, which can be obtained from 138 bags of 50 [kg] or in bulk. On the other hand, the total amount of stone to be used is approximately 29773 [kg] or 29.75 [Ton].

1.1. Plastering.

A plaster of 2 [cm] depth will be considered on the entire support surface of the modules, to ensure that the area exposed by the demolition is uniform and allows the correct support of the blocks.

As the plaster will be exposed to water, a waterproof plaster is designed according to the following volume dosage:

The materials needed for the preparation of the water-repellent plaster are as follows: 3 parts of fine sand, 1 part of cement and the water-repellent which is prepared with 10 litres of clean water and 1 kilo of the water-repellent additive product.

Given that the plaster will be 2 [cm] thick, 20.4 [cm] high (due to the slope of the lateral faces) and 371 [m] long, corresponding to the entire projected work, the volume of plaster for both sides of the module will be 3.03 [m³]. Considering a wastage of 10 %, the final volume will be 3.33 [m].³

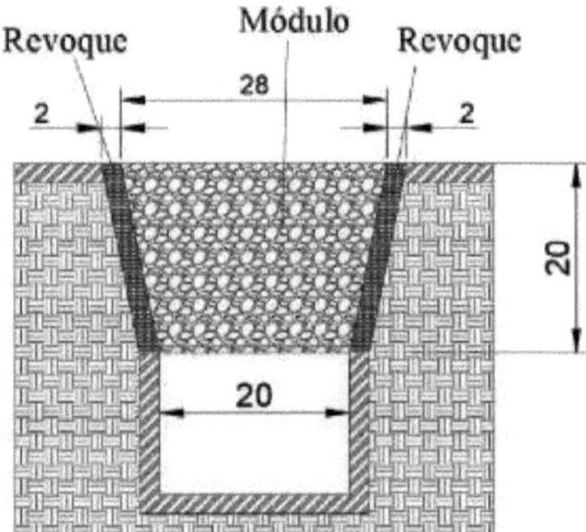

Figure N° 78: Cut showing the plaster.

From the above, the final quantities of materials can be obtained.

Cement: 1172.36 [kg]. 24 bags of 50 [kg].

Sand: 5,978 [m³]. 6 bags of 1 [m].³

Water repellent: 29.89 [kg]. 2 containers of 10 [litres].

2. Demolition.

From the above, it is known that the space currently available for drainage in the pedestrian area to be intervened is 20 [cm] wide, as the module has 28 [cm] on its upper face, part of the pavement will have to be demolished to place the modules and execute the plastering.

Area to be demolished Area to demolish

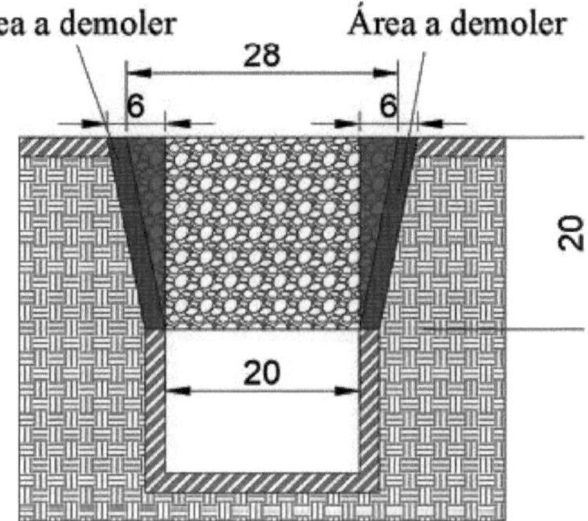

Figure N° 79: Cross-section of the module and area to be demolished (scale in centimetres).

Volume to be demolished:

$$V_d = A_d * L$$

$$A_d = \frac{(6\,[cm] + 2\,[cm])}{2} * 20\,[cm] * 2 = 160\,[cm^2] = 16.10^{-3}[m^2]$$

$$V_d = 16.10^{-3}[m^2] * 371\,[m] = 5{,}936\,[m^3] \approx 6\,[m^3]$$

A total of 6 [m^3] of pavement must be demolished, which can be placed in one 8 [m^3] container or in two 5 [rn^3] containers, the most efficient option being evaluated according to the cost in each case. They must then be correctly unloaded in the place provided by the municipal authorities for this type of waste. The containers will be fitted with the gratings removed from the existing drainage structure.

3. *Time of work and number of workers.*

Demolition:

The tasks considered in this item are the movement of the rubble produced into a container and the actual demolition of the volume of pavement detailed above with a pneumatic rotary hammer, a tool particularly designed for this purpose.

Since the task to be performed requires a certain degree of precision, it is assumed that a worker with his work tool in an 8-hour working day will demolish 0.25 [m^3], or the equivalent of approximately 30 [m] linear metres, together with the task of disposal in the container.

Considering two workers, the total demolition time required will be 6 [days]. Allowing for delays due to reduced efficiency, rainy days, etc., 10 full days are considered to complete the task.

Placement of modules:

Knowing the number of modules to be placed (780), from the limit values for the manual lifting of loads (Res. SRT 295/03), the number of workers needed to place the blocks can be obtained.

These limit values recommend conditions for manual lifting of loads in workplaces, considering that most workers can be exposed repeatedly, day after day, without developing work-related low back and shoulder disorders associated with repeated manual lifting tasks.

The values are contained in three tables with weight limits, in Kilograms [Kg], for two types of load handling (horizontal and overhead), for manual mono-lifting tasks, within 30 degrees of the sagittal (neutral) plane. The limit values are given for manual lifting tasks defined by their duration, whether this is less or more than 2 hours per day, and by their frequency expressed by the number of manual lifts per hour, as defined in the notes to each table. Table 2 of Resolution SRT 295/03, shown in Figure 72, indicates the maximum weights that can be lifted by a worker according to the movement required and the number of lifts per hour.

TLVs for manual lifting for tasks > 2 hours per day with > 12 and < 30 lifts per hour or < 2 hours per day with 60 and < 360 lifts per hour.

TABLA 2. TLVs para el levantamiento manual de cargas para tareas > 2 horas al día con > 12 y ≤ 30 levantamientos por hora o ≤ 2 horas al día con 60 y ≤ 360 levantamientos/hora.

Situación horizontal del levantamiento / Altura del levantamiento	Levantamientos próximos: origen < 30 cm desde el punto medio entre los tobillos	Levantamientos intermedios: origen de 30 a 60 cm desde el punto medio entre los tobillos	Levantamientos alejados: origen > 60 a 80 cm desde el punto medio entre los tobillos[A]
Hasta 30 cm[B] por encima del hombro desde una altura de 8 cm por debajo del mismo.	14 Kg	5 Kg	No se conoce un límite seguro para levantamientos repetidos[C]
Desde la altura de los nudillos[B] hasta por debajo del hombro.	27 Kg	14 Kg	7 Kg
Desde la mitad de la espinilla hasta la altura de los nudillos[D]	16 Kg	11 Kg	5 Kg
Desde el suelo hasta la mitad de la espinilla	14 Kg	No se conoce un límite seguro para levantamientos repetidos[C]	No se conoce un límite seguro para levantamientos repetidos[C]

Figure N° 80: Cross-section of the module and area to be demolished (scale in centimetres) [52].

Corazza Ignacio

In order to lift one of the modules within the maximum weights, the lift must be carried out by at least two people, and if they work a full 8-hour day, they can carry out up to 30 lifts per hour, giving a total of 240 lifts per day.

Since the modules will have to be unloaded from trucks, unloading would be the first survey carried out, assuming that the modules are not stored, but installed directly, the second survey corresponds to the placement of the models in their final position on site.

With 780 modules, a minimum of 1560 lifts must be carried out, two per module, one for unloading and one for installation. Two workers can carry out this task in 6.5 working days. Allowing for delays due to reduced efficiency, rainy days, etc., 10 full working days are considered to complete the task. This would imply two weeks of work from Monday to Friday, which is a reasonable time for the work analysed.

Plastering:

For plastering and rendering in general, skilled labour shall be used. Work crews must be equipped with appropriate trestles and scaffolding. The equipment and tools required shall be in good condition and in sufficient quantity. Screeds shall be made of metal or wood, of suitable cross-section, sharp-edged and straight.

In estimating the number of workers needed to execute the plaster in a given time the plaster is computed per square metre, giving a total of 151.37 $[m].^2$

Considering that a skilled worker will be able to carry out 1 $[m^2]$ per hour of plastering, if two workers are hired to work 8 hours per day, the entire plastering work will be completed in 10 [days].

Summary:

A total of six workers and a foreman will be required for the entire construction site (excluding the transport of modules and containers). As the modules will be materialised in a plant (prefabricated) and transported to the construction site, the man-hours are not counted, as they will be the responsibility of a third party contracted for this part of the work.

The following table summarises the number of days required for each task and the total construction time, it is assumed that after 30 % of the demolition work has been completed, the modules can be installed. After 2 days of setting time for the plaster, the modules can be installed. The calculation takes into account that no work will be carried out on Saturdays and Sundays. This means that 19 working days will be necessary to complete the work.

TaskDays	1	2	3	4	5	6	7	8	9	10	11	12	13	14	15	16	17	18	19
Demolition																			
Plaster																			
Placement of modules																			

Table N° 11: Scheme of working hours required.

4. Machinery required.

Heavy machinery can be dispensed with for unloading or moving the drainage concrete modules, as they are designed to be categorised as lightweight precast elements, so that they can be lifted and placed by two workers without the need for mechanical assistance.

For the plasters, a mixer is considered necessary in order to guarantee the estimated performance, as the volume of mortar is approximately 3 [m^3], which is too large to mix manually due to the time involved in the task.

For the demolition of pavements, pneumatic rotary hammers are needed. These are portable hammer drills that work with compressed air mechanisms, functioning like a hammer, striking the surface with the aim of obtaining pieces of the material to be demolished. These machines are generally used for professional use to make holes in paved floors or to demolish different types of constructions.

Figure N° 81: Operator using a pneumatic rotary hammer [53].

The figure shows that the task carried out is similar to that proposed in the project and the use of personal protective equipment (ear muffs, goggles, gloves, safety shoes, fireproof trousers). As the number of demolition workers is two, the use of two pneumatic rotary hammers is envisaged.

5. *Formwork, manufacture and transport.*

Metal formwork was chosen in order to guarantee uniformity of dimensions for all the materialised modules. These can be manufactured on site or in the plant. The best option is to manufacture the modules in the plant, since it is possible to have greater control over the materials, the dosages and the environmental conditions for curing, as well as to obtain an off-site storage space, allowing the public space occupied during the construction period to be reduced to a minimum, thus reducing the inconvenience for passers-by, shops and neighbours on the pedestrian walkway.

According to what has been studied, the modules can be stripped the day after the draining concrete has been poured into the moulds, and can be stored without load until they acquire the necessary strength, applying a damp curing.

Next, an estimate is made of the amount of metal formwork necessary to guarantee a production that allows the work to progress without delays or interruptions but without incurring unnecessary expenses due to excess moulds, that is to say, production optimisation will be sought.

Considering that 78 modules will be needed each laying day, the production will be divided into 10 batches of 78 produced units. For this, 80 shutterings are needed (2 faulty modules per batch are considered) and to start continuous production 28 days before the first day of laying (so that the material acquires the design strength), arriving at the end of the work with the last blocks produced.

6. *Budget.*

The cost estimate will be made for the manufacture of the modules (materials associated with the draining concrete block), the man hours for the demolition of the volume detailed above, the execution of the plaster and the final placement of the modules. The price of the formwork and the man-hours required for its manufacture will not be taken into account, as this will depend on the company contracted to carry out this task and will be variable according to the internal logistics of each bidder, or another manufacturing configuration may even be considered (wooden or phenolic formwork, on-site manufacture, etc.), which will be evaluated at the time of awarding the project. The quantities of materials, the volume to be demolished and the plaster to be applied remain unchanged. Therefore, only the costs associated with the materials and labour for the aforementioned stages will be detailed, as these are essential for the execution of the work.

Labour:

The cost associated with labour depends on the collective labour agreement of the Argentinean Construction Workers' Union (UOCRA), together with the Argentinean Chamber of Construction (CAMARCO) and the Argentinean Federation of Construction Entities (FAEC), these entities jointly establish the minimum wage to be paid to construction employees, whether they are skilled tradesmen, journeymen or any other position within the activities covered by the agreement [54].

The data are obtained from a table indicating the basic wage rates effective as of 01 December 2022, included in the latest labour agreement for construction workers. These amounts may vary according to future agreements, therefore all amounts indicated in Argentine pesos ($) will be converted to the monetary unit of the US dollar (USD), in order to have a long-term reference parameter associated with costs.

The calculations are based on the table indicating the basic wage effective as of 1 March 2023. The city of Santa Fe belongs to zone "A" [54].

It is considered that the workers hired for plastering and demolition tasks will be specialised journeymen, as they will have to carry out tasks on site with a certain degree of difficulty (precise demolition, inclined plastering and with little working space). The foreman will also be a specialised journeyman. The workers contracted to lay the draining concrete modules will be assistants.

From this we obtain that the minimum hourly wage for a specialised officer will be $948 (nine hundred and forty-eight Argentinean pesos per hour), equivalent to USD 4.83 (four dollars and eighty-three US cents), Banco Nación exchange rate as of 09/02/2023.

For an assistant, the minimum hourly wage will be $620 (six hundred and twenty Argentinean pesos per hour), equivalent to USD 3.16 (three dollars and sixteen US cents), Banco Nación rate as of 09/02/2023.

Table 11 gives the number of days and workers needed to carry out the project, resulting in the following:

For demolition, two specialised officers will be hired to work 8 hours a day for 10 days, with a final salary of $75840 (seventy-eight thousand eight hundred and forty Argentinean pesos) or USD 386.45 (three hundred and eighty-six dollars and forty-five US cents) per officer. Final: $151680 (one hundred and fifty-one thousand six hundred and eighty Argentinean pesos) or USD 772.89 (seven hundred and seventy-two dollars and eighty-nine US cents).

For plastering, two skilled labourers will be hired to work 8 hours a day for 10 days, with a final salary of $75840 (seventy-eight thousand eight hundred and forty Argentinean pesos) or USD 386.45 (three hundred and eighty-six dollars and forty-five US cents) per labourer. Final: $151680 (one hundred and fifty-one thousand six hundred and eighty Argentinean pesos) or USD 772.89 (seven hundred and seventy-two dollars and eighty-nine US cents).

For the placement of the drainage concrete modules, two assistants will be hired to work 8 hours a day for 10 days, with a final salary of $ 49600 (forty-nine thousand six hundred Argentinean pesos) or USD 252.74 (two hundred and fifty-two dollars and seventy-four US cents) per assistant. Final: $ 99200 (ninety-nine thousand two hundred Argentinean pesos) or USD 505.48 (five hundred and five dollars and forty-eight cents).

For a foreman, a specialised officer will be hired to work 8 hours a day for 19 days, with a final salary of $ 144096 (one hundred and forty-four thousand ninety-six Argentinean pesos) or USD 734.25 (seven hundred and thirty-four dollars and twenty-five cents) for the foreman.

This results in total labour costs of $ 546656 (five hundred and forty-six thousand six hundred and fifty-six Argentinean pesos) or USD 2785.51 (two thousand seven hundred and eighty-five dollars and fifty-one US cents).

Materials:

Cement: adding the amount of cement needed for the modules and the cement for the plaster, we obtain a quantity of 6878 *[kg]* + 1172.36 [kg] ≈ 8051 *[κ∂]*. Considering a 10% wastage, 8856 [kg] would be the total kilograms of cement required.

It is planned to use cement in 50 [kg] bags, so at least 78 bags of cement will have to be purchased. The wholesale price (pallet) is currently quoted at $ 88800 (eighty-eight thousand eight hundred Argentinean pesos) or USD 452.48 (four hundred and fifty-two dollars and forty-eight US cents), (value obtained from consultation with wholesalers). Each pallet contains 40 units, therefore 2 pallets will be needed to cover the demand. The final amount of cement will be $177600 (one hundred and seventy-seven thousand six hundred Argentine pesos) or USD 904.97 (nine hundred and four dollars and ninety-seven cents).

Stone: The quantity of stone required for the project is 29773 *[kg],* which will be used in the manufacture of the permeable concrete modules. Bulk purchase is envisaged, i.e. in bulk. The granite split stone has a specific weight of approximately 1600 [^| ·], so 18.61 [m^3] will be needed, considering a 10 % waste, the final value is 20.47 [m].[3]

The price per cubic metre of stone item 6-19 is currently $16900 (sixteen thousand nine hundred Argentinean pesos) or USD 86.11 (eighty-six dollars and eleven cents), (value obtained from consultation with wholesalers).

Since 21 [m^3] of granite 6-19 split stone are to be purchased, the final price is $354900 (thirty-five hundred and fifty-four thousand nine hundred Argentinean pesos) or USD 1808.41 (one thousand eight hundred and eight dollars and forty-one US cents).

Sand: A total of 6 [m^3] of sand will be needed, which will be used for plastering. Considering a wastage of 10 %, 6.6 [m^3] will be needed. As this is a low number, the purchase of 0.75 [m^3] bags is proposed, which also allows for better handling on site and less waste.

Currently the price of a bag of 0.75 [m^3] of fine, clean sand is $ 6900 (six thousand nine hundred Argentine pesos) or USD 35.16 (thirty-five dollars and sixteen cents), (value obtained from consultation with wholesalers).

Since the equivalent of 6.6 [m^3] of fine, clean sand must be purchased, the final price is $62100 (sixty two thousand and one hundred Argentine pesos) or USD 316.43 (three hundred and sixteen dollars and forty three US cents), corresponding to 9 bags of 0.75 [m].[3]

Water-repellent: A total of 29.89 [kg] of water-repellent material will be needed, which will

be used for the plasters. Considering a waste of 10 %, 33 [kg] will be needed. As the number of buckets of 10 [kg] is low, this allows for better handling on site and less waste.
Currently the price of a 10 [kg] bucket of water-repellent material is $ 2704.30 (two thousand seven hundred and four pesos and thirty Argentinean cents) or USD 13.78 (thirteen dollars and seventy-eight US cents), (value obtained from consultation with wholesalers).
Since the equivalent of 33 [kg] of water-repellent material must be purchased, the final price is $ 10817.20 (ten thousand eight hundred and seventeen Argentine pesos and twenty cents) or USD 55.12 (fifty-five dollars and twelve cents), corresponding to 4 buckets of 10 [kg].
Summary:

Post	Quantity	$	USD
Specialised officer	5	447456	2280.03057
Assistant	2	99200	505.48
Total	7	546656	2785.51

Table N° 12: Summary of labour costs.

Material	Quantity	$	USD
Cement	8856[kg]	177600	904.97
Stone	20.47 $[m]^3$	354900	1808.41
Sand	6.6 $[m]^3$	62100	316.43
Water-repellent	33 [kg]	10817.2	55.12
Total	-	594600	3084.93

Table N° 13: Summary of material costs.

Complete works	$	USD
Total	1141256	5870.44

Table N° 13: Final cost of work.

Based on the above, the total cost of the work, taking into account the items indicated, is $ 11411256 (one million one hundred and forty-one thousand two hundred and fifty-six Argentinean pesos) or US $5870.40 (five thousand eight hundred and seventy dollars and forty cents).
Since the work has a length of 371 [m], the cost per linear metre associated with the proposed project is $ 3076.16 (three thousand seventy-six pesos and sixteen Argentinean cents) or USD 15.67 (fifteen dollars and sixty-seven US cents), considering labour and materials. If only materials are taken into account, the final cost per linear metre is $1602.70 (one thousand six hundred and two pesos and seventy Argentinean cents) or USD 8.17 (eight dollars and seventeen US cents).
The percentage of the cost associated with materials is 52%, while labour accounts for the remaining 48%.

7. *Cost comparison.*

A cost comparison is made with the steel gratings currently used as a solution to cover the storm drains of the pedestrian walkway.
From the consultation with entities that can supply the metal grilles, a final quotation per linear metre of $ 17247.65 (seventeen thousand two hundred and forty-seven pesos and sixty-five Argentinean cents) or USD 87.89 (eighty-seven dollars and eighty-nine US cents) was

obtained. Giving a final value for the 371 [m] equal to $ 6398878.15 (six million three hundred and ninety-eight thousand eight hundred and seventy-eight pesos and fifteen Argentinean cents) or USD 32605.75 (thirty-two thousand six hundred and five dollars and seventy-five US cents).

These figures show that savings of up to 90.7 % can be made on materials alone by using drainage concrete modules. Considering all associated costs (including labour), the price is still lower and the percentage saving is 82.16 %. This implies that for every linear metre constructed with the currently used metal gratings, approximately 5.6 linear metres could be constructed with the designed drainage concrete modules.

It is therefore concluded that not only will all the benefits of using drainage concrete be obtained, but also the costs associated with the construction of the drainage inlets will be reduced.

Chapter IX

Risk analysis.

1. Introduction.

In this chapter, the project will be approached from the point of view of the risks to which it will be subjected by the environment and those that it will itself generate towards its location.
In order to identify the type of analysis to be carried out within the framework of risk management, the following diagram shows the stages into which a civil engineering investment project is generally subdivided and the place of this Final Project.

Phases of the PoI	Stages of the phase	Intervention by the GoR
Pre-investment	Idea	Threat analysis and mapping
	Profile	
	Pre-feasibility	**Vulnerability analysis (exposure, fragility and resilience)**
	Feasibility	
	Executive project	Identification of risk reduction measures
Investment	Execution	Implementation of risk reduction measures
Operation	Operation and maintenance	Monitoring and follow-up of risk reduction measures

Table N° 12: Risk management intervention at different stages of an investment project. Source: lecture notes.

In the above table, the cells that identify the stage in which this final degree project is framed are highlighted in green. Accordingly, the risk management analysis will be approached with emphasis on the identification of threats, taking into account those that occur from the environment to the project and from the project to the environment.
On the other hand, vulnerabilities will be analysed in order to evaluate possible improvement interventions to mitigate or eliminate the negative effects produced by the occurrence of the identified hazards. It should be clarified that the objective of this management is not the elimination of risk, but rather an "intelligent coexistence" with it, a situation that is only possible if a correct identification and evaluation of the risks is carried out during the pre-investment phase. This assessment is carried out by carefully observing the project development scenarios, in order to anticipate potential failures of the system with preventive actions and interventions.
The following is a definition of certain terms that are important for the correct understanding of the following analysis.

- **Hazard:** is the existence of any factor, source or situation with the probability of causing social, economic or environmental damage over a given period of time. They can be classified as natural, socio-natural and anthropogenic.

Natural: these are those not produced by man, i.e., in which human activity does not intervene for their existence, such as earthquakes, volcanic eruptions, some types of floods, landslides, among others. They can also be subclassified as follows.
Endogenous: earthquakes, tsunamis, volcanic eruptions.
Exogenous: alluvium, landslides, avalanches, fluvial erosion.
Hydrometeorological: drought, hurricanes, cyclones, frost, river and rain flooding, hail, snowfall.
Socio-natural: hazards caused by natural phenomena conditioned by human activity, e.g.

consequences of climate change, floods due to lack of planning, etc.
Anthropic: hazards that exist exclusively due to human activity, such as leaks and spills of chemical substances, explosions or fires of materials stored or manipulated by man, landslides in excavations, etc. These are all situations that would not be possible without human activity.

- **Vulnerability:** conditions determined by physical, social, economic and environmental factors or processes that increase the susceptibility and exposure of systems and/or people to the impact of hazards.
- **Risk:** the potential for harm to exposed people or systems, depending on the probability of occurrence of hazards and the degree of vulnerability of the system or set of people under analysis.

2. *Description of the analysis.*

The analysis will be carried out on the basis of the following type of table provided by the chair:

Item	Situation			Risk		Improvement intervention	
	Applies	Not applicable	Not defined	Threat	Vulnerability	Action	Improve

Table N° 13: Model table for risk analysis. Source: lecture notes.

The first step is to indicate the preset items that are the subject of the assessment of this methodology. Then, in the "situation" column, it will be marked whether the mentioned item applies to the problem of the present project.
Situations are analysed such as, for example, the noise caused by pneumatic hammers, which can reach 100 decibels at 2 metres, constitutes a risk of hearing loss due to continuous use, the main symptom being tinnitus. In the case of a manual hammer, the operator must wear anti-noise safety earmuffs, and the hours of silence established in urban areas and the maximum decibel limits must be taken into account.
The following abbreviations will be used in the table.

- **V.A.**: High feasibility. The intervention can be carried out in a relatively simple way, with the simple application of security elements, activity planning or other measures.
- **V.M.**: Moderate feasibility. Training and regular monitoring are needed to ensure the efficiency of the intervention.
- **V.B.**: Low feasibility. Intervention requires special studies and does not guarantee significant risk reduction.

The complete tables can be found in the appendix, one for the Management of Risks from the Environment to the Project and another for the Management of Risks from the Project to the Environment.
From the observation of the tables, it can be seen that most of the risk mitigation measures are highly feasible to carry out and represent basic issues that in most cases are taken into consideration when considering projects of this nature. However, there are risks that are difficult to reduce. These are attributed to everyday tasks that are characteristic of this type of works, which can occasionally generate some kind of inconvenience for the social sectors affected by the construction work.
In the same way, a project such as this aims to provide an essential improvement to the quality of life of the local people, so it is possible that the inconveniences generated by the construction work will be understood and accepted by a large part of society.

Chapter X

Environmental impact.

1. Introduction.

The Environmental Impact Assessment (EIA) is the mandatory procedure for identifying, predicting, evaluating and mitigating the potential impacts that a project, work or activity may cause to the environment in the short, medium and long term; it is an instrument that is applied prior to making a decision on the implementation of a project [55].

When applied at a stage prior to the execution of a project, the environmental impact assessment allows the search for alternatives and selects the best one for the fulfilment of the project's objectives, integrating the environmental variable from the beginning. This makes it possible to work on the endogenous variables of the project and not on the exogenous ones, that is, to work on what may cause an environmental impact and mitigate it before it occurs, otherwise only palliative measures can be taken on the impacts already produced or those that will inevitably be generated by the executed project.

The content that must be submitted to the environmental authority in the central technical document of the procedure, called the Environmental Impact Study, is developed below. This study must contain a description of the project, an environmental diagnosis, the legal framework, the identification and assessment of the potential environmental impacts that the project may cause in all its stages and the mitigation measures to address them, the latter of which must be structured in an Environmental Management Plan.

2. Direct and indirect area of influence.

Direct: This is considered to consist of the affected urban area, the population centres that make it up and the space where the work will be located and which during the construction phase will have the site for the construction site and the deposit. The area of direct influence of the project is the sub-basin where the work will be carried out, being the area where the greatest influence will be felt due to the modifications made to the urban drainage system. In addition, neighbours and passers-by will be affected to a greater extent by the construction phase, with noise and interruptions in the pedestrian area.

Figure N° 82: Area of direct influence of the project [13].

Indirect: Corresponds to the physical space in which a directly affected environmental component in turn disturbs one or more other environmental components not directly related to the Project. The area of indirect influence of the project is the entire city of Santa Fe, since all the inhabitants will benefit from the reduction of environmental pollution as a result of the work carried out. They will also benefit from the reduction of costs in the application of the modules. On the other hand, since the project is located in the central area of the city, it is considered that a large percentage of inhabitants frequent the area for various activities, so they will be negatively affected by noise and interruptions in the pedestrian area.

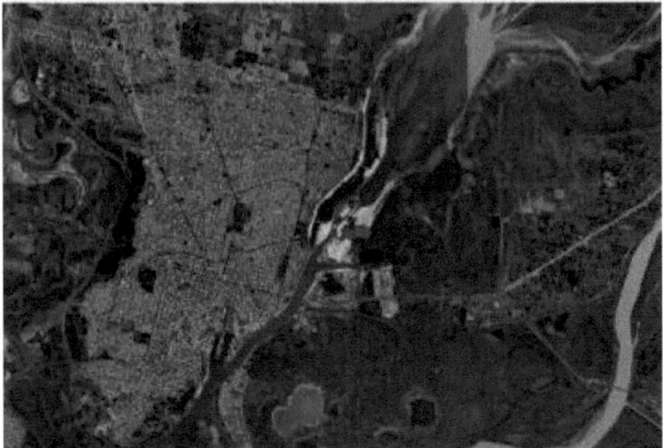

Figure N° 83: Area of indirect influence of the project [56].

3. Legal framework.

The objective of this point in the Environmental Impact Study is to present the environmental legal regulations applicable to the project, according to the type of work or activity, the location and the environmental aspects identified. It also includes the applicable National, Provincial and Municipal regulations based on the location of the project. It also contains the international treaties on environmental matters that have been adopted by the country and must be considered as appropriate.

All articles of the National Constitution, national and provincial laws and municipal ordinances with environmental relevance shall be indicated with a brief description of what they indicate.

National Constitution: [57].

Article 41: all inhabitants have the right to a healthy, balanced and suitable environment for human development and for productive activities to satisfy present needs without compromising those of future generations; and they have the duty to preserve it.

Article 43: Any person may bring a prompt and expeditious action for amparo, provided that there is no other more suitable legal remedy, against any act or omission of public authorities or private individuals that actually or imminently injures, restricts, alters or threatens, with manifest arbitrariness or illegality, rights and guarantees recognised by this Constitution, a treaty or a law.

National laws: [58] (The laws that are considered to be the main ones for the project are listed).

Law N° 25.675, "General Environmental Law": establishes the minimum requirements for the achievement of a sustainable and adequate management of the environment, the preservation and protection of biological diversity and the implementation of sustainable development. Argentine environmental policy is subject to compliance with the following principles: congruence, prevention, precautionary, intergenerational equity, progressiveness, responsibility, subsidiarity, sustainability, solidarity and cooperation.

Law No. 25.831: Guarantees the right of access to environmental information held by the State, at the national, provincial, municipal and Autonomous City of Buenos Aires levels, as well as by autonomous entities and public service providers, whether public, private or mixed.

Law No. 25.612: Regulates the integrated management of industrial waste and waste from service activities, which are generated throughout the national territory, and are derived from industrial processes or service activities.
Law No. 25.670: Systematises the management and elimination of PCBs, in the whole territory of the Nation under the terms of art. 41 of the National Constitution. It prohibits the installation of equipment containing PCBs and the import and entry into the national territory of PCBs or equipment containing PCBs.
Law N° 25.688: Establishes the minimum environmental budgets for the preservation of water, its exploitation and rational use. For interjurisdictional basins, water basin committees are created.
Law N° 25.916: Regulates the management of household waste.
Law No. 20.284: The purpose of which is to prevent atmospheric pollution, it establishes rules to be applied to all sources capable of producing atmospheric pollution located in federal jurisdiction and in the jurisdiction of the provinces that adhere to it.
Law No. 20.284: Its objective is to structure and implement a national programme involving all aspects related to the causes, effects, scope and methods of prevention and control of atmospheric pollution.
Law No. 19.587: Modifying and complementary norms regulate measures aimed at preserving the psychophysical integrity of workers, in order to reduce accidents and occupational diseases, as well as the risks arising from different factors of labour activity.
Law No. 24.557: Modifying and complementary rules, they make up the regulatory framework that establishes the new comprehensive system for the prevention of occupational risks (SIPRIT), and the legal regime of occupational risk insurers (ART).
Provincial laws: [59] (The laws considered to be the main ones for the project are listed).
Law N° 11.717: Establishes within the integral development policy of the Province, the guiding principles to preserve, conserve, improve and recover the environment, the natural resources and the quality of life of the population. To ensure the inalienable right of every person to enjoy a healthy, ecologically balanced and adequate environment for the development of life and the dignity of the human being.
Law N° 10.000: Establishes the administrative appeal against any decision, act or omission, which, violating current dispositions, harm the simple or diffuse interests of the inhabitants of the province of Santa Fe in the protection of public health, in the conservation of fauna, flora and landscape, in the protection of the environment, in the preservation of historical, cultural and artistic heritage, in the correct commercialisation of goods to the population and, in general, in the defence of similar values of the population.
Law N° 11.872: Prohibits weeding by means of fire, the installation of any type of open-air waste deposit, public or private, the generation of smoke or gases that could cause risks to traffic on provincial and national roads and railways, without being treated with techniques that prevent these consequences.
Law No. 11.717: Substitutes the regulation of articles 22 and 23 of Law No. 11.717 that creates the State Secretariat of Environment and Sustainable Development. It regulates aspects related to the following subjects: registers of consultants; of generators, operators and transporters; of offenders; it establishes rules related to documentation, namely: Manifest; Certificate of Environmental Aptitude, Additional annual fee for generation and operation of hazardous waste; administrative fee for hazardous waste transporters.
Law No. 11.220: Regulatory framework applicable to the provision of drinking water and

sewage services. It determines that industrial effluents must comply with the quality standards, concentration of substances and volumes contained in Annex B of Law 11.220.

Municipal Ordinances: [60] (The ordinances that are considered to be the main ones for the project are listed).

Ordinance N° 11.017: ENVIRONMENTAL IMPACT ASSESSMENT. Environmental Impact Study is understood as the technical administrative procedure destined to identify and interpret, as well as to prevent the short, medium and long term effects that activities, projects, programmes or public or private undertakings may cause to the environment. Environmental Impact is understood as any net change, positive or negative, that is provoked on the environment as a direct or indirect consequence of anthropic actions that may produce alterations susceptible to affect health and quality of life, the productive capacity of natural resources and essential ecological processes.

4. Environmental diagnosis.

Field surveys were carried out to determine the current conditions of the site to be intervened and problems of waste accumulation in the storm drainage pipes were observed, which initiated the project proposal. The aforementioned survey can be seen in the following image.

Figure N° 84: Urban waste in storm drains on Santa Fe's pedestrian walkway.

5. Analysis of environmental impacts.

The Environmental Impact Assessment makes it possible to identify, analyse and describe the impacts that the project will have on its surroundings (physical, biological and socio-cultural-economic environment).

The procedure to be used for the verification of an interaction between the cause (action considered) and its effect on the environment (environmental factors), will be materialised by elaborating Interaction and Identification of Environmental Impacts Matrices, in which each cross cell between cause and effect represents a possible impact for each stage of the project.

5.1. Methodology.

The methodology consists of the identification and evaluation of the interactions of the environmental component in relation to the project activities, according to the stages of construction, operation and closure, in order to subsequently obtain a qualitative-quantitative assessment of the impacts, through parameters of importance and magnitude, and finally categorise the environmental impacts into highly significant, significant, negligible and positive.

For the elaboration of the Matrix of Identification of Environmental Impacts of the project, a Leopold Matrix will be made, which considers in the columns (actions of the works) and in the rows the environmental factors. In this way the whole study area is presented, in order to start the identification of the character of the environmental impact in positive or negative (+/-). Where an impact is expected, the appropriate cell of the matrix is divided diagonally from the top right corner to the bottom left corner, to place the magnitude and importance for each interaction. Magnitude has values from -10 to +10 and is placed in the upper left corner, while importance takes values from 1 to 10 and is inserted in the lower right corner.

Measures of magnitude and significance tend to be related, but are not necessarily directly correlated. Magnitude can be measured more tangibly by considering the area affected by the development or how severe it will be, while significance is a subjective measure. While a proposed project may have a large impact in terms of magnitude, the effects caused may not actually be significant to the environment.

Although the methodology involves the analysis of 8800 interactions between actions and factors that may have impacts, a summarised version will be developed that will consider the main actions of the construction phase, taking into account the factors that are usually involved in similar projects and for which sufficient information is available to analyse them.

5.1.1. Rating of impacts.

Each environmental impact is rated according to its level of importance and magnitude, and according to its sign, positive or negative. In order to generalise these criteria, it has been decided to perform a geometric mean of the multiplication of the values of importance and magnitude. The result of this operation is called the Impact Value and corresponds to the following equation:

$$\text{Impact } value = \pm(\text{Significance x Magnitude})^{0,5}$$

Therefore, an environmental impact can reach an Impact Value between a maximum of 10 and a minimum of 1. The categorisation of the assessed environmental impacts will be done on the basis of the Impact Value by considering four impact categories with their respective identifying colours:

Category	Colour
Highly significant	
Significant	
Despicable	
Positives	

Table 14: Impact categories.

Highly significant impacts: Those of a negative nature, whose "Impact Value" is greater than or equal to 6.5, correspond to the effects of high incidence on the environmental factor, difficult to correct, of generalised extension, with an irreversible type of effect and of permanent duration.

Significant impacts: Those of a negative nature, whose Impact Value is less than 6.5 but greater than or equal to 4.5, whose characteristics are, feasible to correct, of local extension and temporary duration.

Negligible: These correspond to all impacts of a negative nature, with an impact value of less than 4.5. This category includes impacts that can be corrected during the implementation of the Environmental Management Plan, and are characterised by being reversible, sporadic in duration and with a one-off influence.

Positive: These are those of a positive nature that are beneficial, advantageous, positive or favourable produced during the execution of the project and that contribute to the promotion of the project, without causing damage to the environment.

The matrix can be found in the annex, table 17.

As a result of the Leopold matrix, the results can be summarised in the following graph.

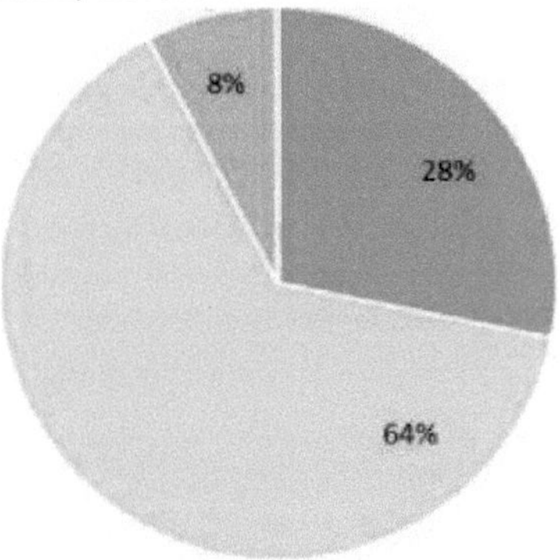

Table N° 85: Graph of EIA results.

It can be seen that the project in its construction phase generates mostly negligible negative impacts.

On the other hand, the most negatively affected environmental factors are those corresponding to the physical environment, such as demolition, masonry work and waste generation. On the other hand, the socio-economic environment is affected by the negative effects on accessibility to homes, businesses and normal traffic within the intervened area.

The positive effects are on flora and fauna, as the project reduces environmental pollution. There are also benefits in terms of employment generation, in terms of public services due to less waste in the rainwater pipes and in terms of the landscape, as it is considered that the draining concrete modules have a greater uniformity of colour and materiality with the materials that currently make up the pedestrian surface, compared to the metal grids.

Environmental factors with the highest level of negative environmental impact:

Noise and vibration: the use of demolition machinery can generate a temporary increase in noise levels, which will affect workers, neighbours and passers-by.

Particulate matter: Demolition and masonry work can produce significant amounts of dust that will be temporarily suspended in the air, affecting workers, neighbours and passers-by. The waste generated can also generate this type of particulate matter.

Chapter XI: Conclusions and recommendations.

In the introduction of the present work, the perceived problem situation was detailed, from this point the starting point for the proposal of the objectives of the project, possible solutions and by means of the logical framework scheme to arrive at an optimal solution for the problem posed, which comprehensively covers the situation and fulfils the objectives. Therefore, it should not be lost sight of the fact that the proposed solution is one of many possible solutions and should not be taken as absolute.

In order to detail the conclusions of the work, it is necessary to recall the problem to which this work was intended to provide a response, which manifested the deficit in water management in urban environments, indicating as an area to be addressed the increase in waste in rainwater (figure 15). Based on this, the construction of structures formed by draining concrete elements at the entrance to urban drains was proposed, with the aim of contributing to integrated urban water management, reducing the urban waste that enters the sewers, i.e. filtering the water before it enters the rainwater conduits. This water is then discharged into natural watercourses without any type of treatment, thus reducing the rate of environmental pollution associated with rainwater.

From this, it is possible to observe the progress made, which will help to respond to one of the biggest problems facing the world's society today, environmental pollution.

First of all, it was demonstrated that both the material and the designed module have sufficient permeability to discharge the flow produced by the design storm and the area that contributes to the drains of the pedestrian walkway. This is essential, since if only the filtration capacity is observed, a water accumulation phenomenon could occur and generate flooding.

In addition, a numerical model of permeable concrete subject to two-dimensional flow in a partially saturated permeable medium was generated, using coefficients obtained from experimental tests, in order to simulate the flow path and velocity inside the material and to corroborate that the flow can flow correctly.

The values indicate the high infiltration capacity of permeable concretes, which allows them to be used in urban drainage components, such as those proposed or other elements of greater geometric complexity that can also be studied using this methodology, making them a plausible tool for design.

It was then demonstrated that the material has the necessary filtration capacity to retain urban waste from the water entering the drains, as its porous structure allows water to enter the drains but waste larger than 16 [mm] to remain on the surface. This is the main point of the project, from which it is concluded that the use of the modules as proposed will reduce the amount of waste entering the storm drains, thus reducing blockages in the sewers, increasing their durability and, finally, reducing pollution of the natural watercourse that receives untreated excess rainwater.

Thirdly, the module was verified mechanically, as it would be subjected to the passage of light vehicles. For this purpose, a computational model was made, where all the mechanical characteristics of the material to be analysed were loaded with the geometry of the designed structure and the configuration of loads that the module would be subjected to. The structure responded satisfactorily and with a wide margin of resistance, thus proving that the module will resist correctly.

It should be noted that after a calculation of the materials and labour required for the projected work, it was concluded that the costs associated with carrying out the projected work are lower than the alternative currently used with metal gratings. The figures showed that savings of up to 90.7 % in materials alone can be achieved by using the drainage concrete modules and that considering all costs (including labour), the percentage saving is 82.16 %. This indicates that for every linear metre constructed with the metal gratings currently used, approximately 5.6 linear metres could be constructed with the draining concrete modules designed in this work.

Recommendations:

This section will specify the recommended future studies and research to be carried out to ensure the correct performance of the designed structure and to improve the understanding of its operation.

First of all, it is suggested that tests be carried out to evaluate the clogging rate of the modules, using representative samples of the fine aggregate and the characteristic residues (most present) in the Santa Fe pedestrian walkway. This will allow a more accurate estimation of the durability of the module in terms of its permeability and to foresee cleaning periods in accordance with the results.

Secondly, the importance of performing 3D flow simulations with data collected from computed tomography scans is pointed out. This information can be used to generate three-dimensional volumes of the internal structure of the material and from this observe the behaviour of the fluid in the matrix of the draining concrete. Within these studies, a clogging simulation can also be incorporated, allowing the areas where clogging would occur at the highest speed to be perceived.

Regarding the functioning of the structure, it is recommended to monitor the modules once they are installed, in order to corroborate the variation of the permeability over time and if there is considerable deterioration of the material.

References

↑ **[1]** United Nations, Department of Economic and Social Affairs, Population Division (2022). World Population Prospects 2022, Online Edition.

↑ **[2]** United Nations. World Urbanization Prospects: The 2018 Revision (2018). Online: https://population.un.org/wup/Publications/Files/WUP2018-KeyFacts.pdf

↑ **[3]** LIYA ESHETU ABERA, (2015). Analysis of Pervious Concrete as a Stormwater Management Tool Using Swmm Modeling. University of Mississippi.

↑ **[4]** ALEJANDRO SECCHI, ROSANA MAZZÓN, (2015). Regulation of stormwater surpluses in urban watersheds.

↑ **[5]** JHA ABHAS, JESSICA LAMOND, ROBIN BLOCH, (2011). Five Feet High and Rising: Cities and Flooding in the 21st Century. The World Bank. East Asia and Pacific Region. Transport, Energy & Urban Sustainable Development Unit.

↑**[6]** El Litoral newspaper. "Alarming pollution in the Setúbal lagoon". Área metropolitana. https://www.ellitoral.com/area-metropolitana/alarmante-contaminacion-laguna-setubal 0 1OWc4TnBQ4.html

↑ **[7]** J. LAMOND, N. BHATTACHARYA, R. BLOCH (2012). The role of solid waste management as a response to urban flood risk in developing countries, a case study analysis.

↑ **[8]** MARIO LEANDRO CASTRO ESPINOSA, (2011). Permeable pavements as an alternative for urban drainage. Pontificia Universidad Javeriana, Bogotá Colombia.

↑ **[9]** YU Kongjian; LI Dihua; YUAN Hong; FU Wei; QIAO Qing; WANG Sisi (2015). "Sponge city: Theory and practice.

↑ **[10]** THU THUY NGUYEN, HUU HAO NGO, (2018). Implementation of a specific urban water management - Sponge City. University of Technology Sydney, Harbin Institute of Technology and Xi'an University of Architecture and Technology.

↑ **[11]** INA-MCSF AGREEMENT (2015). Complementary Act No. 13, Final Report. Study of critical areas due to frequent flooding in the city of Santa Fe. Final Report.

↑ **[12]** AGUIRRE DIEGO, ARGENTO ROMINA, (2021). Drainage H° as a regulator of pluvial surpluses. National Technological University, Santa Fe Regional Faculty.

↑ **[13]** Complete updated plan (2019). Department of Engineering and Projects. Directorate of Engineering. Secretariat of Water Affairs and Risk Management. City of Santa Fe.

↑ **[14]** El Litoral Newspaper. "La Municipalidad instaló una estación de bombeo en avenida Freyre y Catamarca". Área metropolitana.https://www.ellitoral.com/area-metropolitana/municipalidad- instalo-estacion-bombeo-station-avenida-freyre-catamarca 0 xpCem0YXYZ.html

↑ **[15]** Santa Fe (Argentina): Departaments and Localities - population Statistics, Charts and Maps. http://www.citypopulation.de/php/argentina-santafe.php

↑ **[16]** MINISTRY OF HEALTH AND ENVIRONMENT; SECRETARIAT OF ENVIRONMENT AND SUSTAINABLE DEVELOPMENT. Argentine Republic (2005). National strategy for the integrated management of solid urban waste.

↑ **[17]** SAATY, THOMAS; MCGRAW HILL (1998). The Analytical Hierarchy Process.

↑ **[18]** XIN GUAN, JIAYU WANG, FEIPENG XIAO (2021). Sponge city strategy and application of pavement materials in sponge city. Key Laboratory of Road and Traffic Engineering of Ministry of Education, Tongji University, Shanghai, 201804, China.

↑ **[19]** El Litoral newspaper. "Atan con cadenas las bocas de tormenta para evitar que las sigan las robando". Santa Fe City, metropolitan area. https://www.ellitoral.com/area-metropolitana/atan-cadenas-bocas-tormenta-evitar-sigan-robando 0 oXIWOU4oE9.html

↑ **[20]** Ph.D. KARTHIK H. OBLA (2010). Pervious concrete - An overview. The Indian Concrete Journal.

↑ **[21]** ENG. DANIEL PÉREZ RAMOS (2005). Experimental study of permeable concretes with andesitic aggregates. Master's and PhD Program in Engineering, Faculty of Engineering, National Autonomous University of Mexico. Mexico, D.F.

↑ **[22]** KLEIN, N. (2019) Offenporiger Beton. *Lärmarme Straßenoberflächen, Spezialbetone*. Technische Universität München.

↑ **[23]** AMERICAN CONCRETE INSTITUTE. ACI 522R-10 (2011). Report on Pervious Concrete.

↑ **[24]** DESAI, DHAWAL. (2014). Pervious Concrete-Effect of Material Proportions on Porosity. Civil Engineering Portal Diperoleh, 22.

↑ **[25]** Heidelbergbeton. Der offenporige Beton - versickerungsfähig und schallasbsorbierend. Pilot projekt einer Werksstraße in Dränbetonbauweise. https://www.heidelbergcement.de/de/beton/pervacrete

↑ **[26]** WILLIAM R. SELBIG. (2019). Evaluating the potential benefits of permeable pavement on the quantity and quality of stormwater runoff. United States Geological Survey.

↑ **[27]** M. HARSHAVARTHANABALAJI, M. R. AMARNAATH, R.A.KAVIN, S. JAYA PRADEEP (2015). Design of eco friendly pervious concrete. International Journal of Civil Engineering and Technology (IjCIET).

↑ **[28]** VAHID ALIMOHAMMADI, MEHDI MAGHFOURI (2021). Stormwater Runoff Treatment Using Pervious Concrete Modified with Various Nanomaterials: A Comprehensive Review. Sustainability 2021, 13, 8552. https://doi.org/10.3390/su13158552.

↑ **[29]** DAN HUFFMAN (2008). Pervious Pavement. National Ready Mixed Concrete Association NRMCA.

↑ **[30]** ANTHONY TORRES, jIONG HU, AMY RAMOS (2015). The effect of the cementitious paste thickness on the performance of pervious concrete.www.elsevier.com/locate/conbuildmat

↑ **[31]** ENG. FERNANDO IMAZ, ENG. IVAN SORBA (2018). Risk in Construction Activities, Ergonomics - Manual Handling of Loads. Civil Engineering. UTN FRSF.

↑ **[32]** National Water Institute, Semi-Arid Region Central Branch. "Lluvias de Diseño". https://www.ina.gov.ar/cirsa/hidrologia/pdf/INA CIRSA Lluvias de Diseno.pdf ↑ **[33]** Memorandum N° 010/ 2017. Province of Santa Fe. Ministry of Infrastructure and Transport.

↑ **34]** Tables, curves and graphs of the Ministry of Infrastructure and Transport of the Province of Santa Fe (Period 1965 - 2000).

↑ **[35]** PIERRE HORGUE, JAQUES FRANC, ROMAIN GUIBERT, GÉRALD DEBENEST. (2015) An extension of the open-source POROUSMULTIPHASEFOAM toolbox dedicated to groundwater flows solving the Richard's equation. INPT, UPS, IMFT (Institut de Mécanique des Fluides de Toulouse), Université de Toulouse.

↑ **[36]** Prof. Dr.-Ing. KAI-UWE BLETZINGER (2019). Einführung in die Finite-Elemente-Methode. Technische Universität München.

↑ **[37]** https://www.openfoam.com/

↑ **[38]** ISO/ IEC/ IEEE International Standard-Systems and software engineering. ISO/ IEC/ IEEE 24765:2010(E). pp. vol., no., pp.1-418.

↑ **[39]** IALY RAYANE de AGUIAR COSTA, ARTUR PAIVA COUTINHO (2020). Sensitivity of hydrodynamic parameters in the simulation of wáter transfer processes in a permeable Pavement. Federal University of Pernambuco, Recife and Caruaru, PE, Brazil.

↑ **[40]** IGNACIO CORAZZA (2022). Hydraulic performance and urban waste filtration capacity of a concrete drainage module for urban storm drain inlets. Jornada de Investigadores Tecnológicos 2022, Grupo de Investigación en Métodos Numéricos en Ingeniería.

↑ **[41]** QIONG LIU, LIANG LI, LARS VABBERSGAARD ANDERSEN, MIN WU. (2022). Stuying the abrasion damage of concrete for Hydraulic structures under various flow conditions. Cement and Concrete Composites. www.elsevier.com/locate/cemconcomp

↑ **[42]** JIONG ZHANG, GUODONG MA, RUIPING MING, XINZHU ANG CUI, LI LI, HUINING XU. (2017). Numerical study on seepage flow in pervious concrete based on 3D CT imaging. www.elsevier.com/locate/conbuildmat

↑ **[43]** COULSON, J. M., J. F. RICHARDSON, J. R. BACKHURST; J. H. HARKER. (2003). Chemical engineering: basic operations "Chapter 9: Filtration".

↑ **[44]** VANESSA PUDERBACH, KILIAN SCHMIDT, SERGIY ANTONYUK (2021). A Coupled CFD-DEM Model for Resolved Simulation of Filter Cake Formation during SolidLiquid Separation. Institute of Particle Process Engineering, Technische Universität Kaiserslautern.

↑ **[45]** KARA ROGERS (2022). Great Pacific Garbage Patch. Britannica.

↑ **[46]** MARIA EUGENIA GARAT, GUSTAVO ROBERTO LARENZE, ALBERTO JOSE PALACIO, JORGE DANIEL SOTA (2019). Hydrological Performance and Physical-Mechanical Properties of Porous Concretes Elaborated with Aggregates from the Province of Entre Ríos. Grupo de Investigación en Ingeniería Civil, Materiales y Ambiente (GIICMA) Universidad Tecnológica Nacional - Facultad Regional Concordia.

↑ **[47]** KIA ALALEA, HONG S. WONG, C.R. CHEESEMAN. (2017). Clogging in permeable concrete: A review. Journal of Environmental Management.

↑ **[48]** GHUFRAN H. FAISAL, ALI J. JAEEL, THAAR S. AL-GASHAM. (2020). BOD and COD reduction using Porous concrete pavements. Civil Engineering Department, Wasit University, Wasit, Iraq.

↑ **[49]** U.S. Department of Transportation. National Highway Traffic Safety Administration. DOT HS 810 561. The Pneumatic Tire.

↑ **[50]** DANIEL GARCIA, POZUELO RAMOS (2008). Pneumatic-road contact model at low speed. Carlos III University of Madrid.

↑ **[51]** American Association of State and Transportation Highway Officials. AASHTO GUIDE FOR Design of Pavement Structures (1993).

↑ **[52]** MINISTRY OF LABOUR, EMPLOYMENT AND SOCIAL SECURITY (2003). Ministry of Labour, Employment and Social Security. HYGIENE AND SAFETY AT WORK. Resolution 295/2003.

↑ **[53]** Mechanical hammer. Installation of Road Signs and Signals in Pereira - Colombia. 21 December 2008. https://es.wikipedia.org/wiki/Martillo_mec%C3%A1nico

↑ **[54]** Collective labour agreement between the Argentinean Construction Workers' Union (UOCRA), the Argentinean Chamber of Construction (CAMARCO) and the Argentinean Federation of Construction Entities (FAEC) (December 2022).

↑ **[55]** Environmental impact assessment. Ministerio de Ambiente y Desarrollo Sostenible de la Nacion. https://www.argentina.gob.ar/ambiente/desarrollo-sostenible/evaluacion-ambiental/evaluacion-de-impacto-ambiental.

↑ [56]https://ide.ign.gob.ar/portal/apps/Profile/index.html?appid=89b630dced514e8a922d0f9957ce925a

↑ 57] Ministry of Justice and Human Rights. Presidency of the Nation. Legislative Information. http://servicios.infoleg.gob.ar/infolegInternet/anexos/0-4999/804/norma.htm
↑ **[58]** Ministry of Justice and Human Rights. https://www.argentina.gob.ar/normativa
↑ **[59]** Santa Fe Province. https://www.santafe.gov.ar/normativa/
↑ **[60]** Santa Fe Council. Ordinances. https://www.concejosantafe. gov.ar/ordinances/

Annex

Item	Situation			Risk		Improvement intervention	
	Applies	Not applicable	Not defined	Threat	Vulnerability	Action	Improve
Physical and terrain	✓	-	-	Floods	Slope of the city tends to be horizontal. Sufficient runoff capacity to cope with excess rainfall.	Ensure the free flow of water into the pipeline.	V.A.
				Landslide	There are no geological formations such as hills, ranges or mountains nearby. No excavation work is carried out.	-	V.A.
				Damage to equipment	The work does not require the use of large equipment or long usage times.	Consider the purchase of spare tools so as not to interrupt the progress of the work.	V.A.
				Accidents at work	Workers do not work at heights. Open drainage (without grating) is considered a risk factor. Noise produced by the equipment used.	Place temporary protections on the open drain until the final placement of the modules. Use personal protective equipment.	V.A.
				Thefts	The work is being carried out on public roads.	Have security guards on duty during non-working hours and on non-working days.	V.A.
Economic and financial	✓	-	-	Availability of funds	Payment of work certificates depend on the State.	The state should evaluate a financing plan with lenders.	V.A.
				Inflation	It depends on the economic policies of the national state.	Budget provision. Having assets to safeguard value.	V.A.
				Suspension of certificate payments	Construction companies with low financial capacity.	Weighing the financial capacity of bidders in the analysis of the award of the work.	V.A.
				Changes in raw material prices	Cement price dependent on the global market.	Generate stock of basic materials.	V.A.
Social	-	-	✓	-	-	-	-
Politician	✓	-	-	Workers' strikes.	Progress of works conditioned by the work of the workers.	Adhere to the Collective Bargaining Agreement in force.	V.A.
Technician	✓	-	-	Faulty project design	Difficulty of reformulation at the construction stage.	Evaluate different design alternatives	V.A.
				Errors in technical specifications	Contractor's work dependent on technical specifications	Generate instances of control of the drafting of specifications.	V.M.
				Deficits in basic education	Efficient design dependent on basic studies	Conduct basic field studies and background research.	V.A.
				Poor on-site progress	Payment of certificates according to the progress of the work	Study and optimise resources to make progress on several simultaneous fronts.	V.A.
				Inexperience of the contractor	Project with special works and specific items of work	Weighing the technical background of the bidders in the analysis of the award of the work.	V.A.
Cultural	-	✓	-	-	-	-	-
Ecological	-	✓	-	-	-	-	-
Institutional	-	-	✓	-	-	-	-

Table N° 15: Assessment of risks of the physical and social environment to the Investment Project.
Corazza Ignacio

Item	Situation			Risk		Improvement intervention	
	Applies	Not applicable	Not defined	Threat	Vulnerability	Action	Improve
Physical and terrain	✓	-	-	Surface storage of waste.	Highly trafficked commercial area in need of cleaning.	Regular sweeping and cleaning.	V.A.
				Affectation of public service infrastructures.	Clogging of the module.	Periodic backwash cleaning.	V.M.
Economic and financial		✓	-	-	-	-	-
Social	✓	-	-	Disruption of pedestrian flow	Highly trafficked pedestrian street.	Delimit the working area to the minimum necessary area.	V.A.
				Increased probability of occurrence of claims due to the construction site.	People passing through areas near the construction site.	Fencing and signposting of the work area.	V.A.
				Cessation of consumer activities in commercial sectors.	The work is being carried out in the main shopping area of the city of Santa Fe.	Delimit the working area to the minimum necessary area. If necessary, plan safe temporary access and crossings to the shops.	V.A.
Politician	-	✓	-	-	-	-	-
Technician	-	✓	-	-	-	-	-
Cultural	-	-	✓	-	-	-	-
Ecological	✓	-	-	Increased emissions of noise and pollutants into the atmosphere.	Work with equipment that emits disturbing noises, with pneumatic hammers being the most noisy.	Alternation in the use of equipment, in order to achieve spaced periods of time without the emission of undesirable noise, respecting the hours of silence and the decibel limits established by the municipality.	V.A.
				Generation of debris and other solid waste.	The work involves stages of demolition, and there is also waste from the materials used.	Waste management planning. Conduct an environmental impact assessment.	V.A.
				Use of cement.	People passing through areas near the construction site who may breathe in the particles expelled. Energy consumed and polluting gases released into the atmosphere for its production.	Increase the efficiency of work involving cement, keep the site clean and tidy. Avoid air pollution with cement by using protective cloths and covers on the mixing machine and if necessary by surrounding the work area. Waste management planning. Conduct an environmental impact assessment.	V.M.
Institutional	-	-	✓	-	-	-	-

Table N° 16: Assessment of risks of the Investment Project to the physical and social environment.
Corazza Ignacio

X. Activities			Collection of materials				Demolition				Masonry				Placement of modules				Movement of vehicles and machinery				Waste generation			
Environmental components			±	M	I	V	±	M	I	V	±	M	I	V	±	M	I	V	±	M	I	V	±	M	I	V
Physical environment	Soils					-	-	2	7	3.7	-	1	7	3.7				-				-	-	2	9	4.2
	Water					-	-	2	7	3.7	-	1	7	3.7				-				-	-	2	9	4.2
	Air	Noise and vibration				-	-	5	4	4.5								-	-	2	4	2.8				-
		Particulate matter	-	3	5	3.9	-	7	5	5.9	-	3	5	3.9				-				-	-	7	5	5.9
		Odours				-				-				-				-				-				-
Natural environment	Vegetation					-				-				-	+	5	8	6.3				-	-	1	8	2.8
	Fauna					-				-				-	+	5	8	6.3				-	-	1	8	2.8
Socio-economic background	Population					-	-	3	6	4.2				-				-				-	-	1	7	2.6
	Health and safety	Labour	-	2	6	3.5	-	2	6	3.5	-	4	7	3.5	-	1	7	2.6				-				-
		Public	-							-				-	+	5	7	5.9				-	-	2	7	3.7
	Economy	Employment	+	3	7	4.6	+	4	7	5.3	+	4	7	5.3	+	4	7	5.3	+	5	7	5.9				-
		Commercial activity				-	-	4	4	4.0				-	+	3	7	4.6				-				-
	Accessibility		-	2	7	3.7	-	2	7	3.7	-	2	7	3.7	-	2	7	3.7				-				-
	Public services					-				-				-	+	5	5	5.0				-	-	2	5	3.2
	Landscape					-	-	1	1	1.0				-	+	3	3	3.0				-				-

Table N° 17: Leopold matrix.

Printed by Books on Demand GmbH, Norderstedt / Germany